BEFORE THE BEGINNING
...God Designed

Jim Kraft

BOOKS

Waite Hill, OH 44094

First Blossom Ridge Books printing, January 2013
www.blossomridgebooks.com

Copyright © 2011, 2012, 2013 by James Kraft. All rights reserved. No part of this book may be used or reproduced in any matter whatsoever without written permission of the publisher except in the case of brief quotations in articles and reviews. For information, write Blossom Ridge Books, 6900 Eagle Mills Road, Waite Hill, OH 44094.

ISBN 13: 978-0-9884652-3-7
eBook ISBN 13: 978-0-9884652-0-6
Corresponding Discussion Guide ISBN 13: 978-0-9884652-2-0

Library of Congress Control Number: 2012919072

Cover Design: Jim Kraft

Editor: Linda Cizek
Assistant Editor: Tamara Kraft

Background Illustration: Painted by Achim Prill,
PRILL Mediendesign & Fotografie
Foreground Photo Illustration: Jim Kraft
Back Cover Photo: The Visual MD

Scripture quotes are from New International Version and New American Standard versions of the Bible.

Book Production: Scribenet

Animal, leaf, and DNA/Fish Emblem Photo Illustrations: Jim Kraft
DNA Strand Illustration: Hakusan
Giraffe Photos: Fotia.com
Woodpecker Photo: Andrei Stroe
Incubator Bird Photo: J.J. Harrison
Owl Butterfly Photo: Didier Descouens
Monkey Face Orchid Photo: Butterfield

For information regarding author interviews,
please contact the publicity department at 440-946-5005.

Printed in the United States of America

Dedicated to Mom, Mike Murray and to those who find themselves asking, What's this all about anyway? And to those who enjoy epiphanies!

Contents

Acknowledgements		ix
Preface		xi
1	**What Was He Thinking?**	**1**
	Awesome Minds	12
	Loving Us Through Design	14
	Passion	15
2	**God's Heart in Communication With Man's Heart**	**23**
	The Ultimate Design Challenge!	33
3	**Inspirations of God**	**37**
	God's Heart	43
	Passion in Action	47
	Really Get It!	49
	Perspective	50
4	**God's Creation: Get the Message**	**53**
	Considering Creation	57
	Humor	58
	Extremes	64
	Awe Factor	67
	Variety	71
	Everyone	76
	Nuts and Bolts Practicality	78
5	**Diversity and Devotion**	**83**
	Devotion	87
6	**The Garden Around Us**	**93**
	Care and Provision	100

	Sweat of the Brow	102
	Perfection	105
7	1.2 oz.–2,000 lb. Beasts	107
	Animals That Challenge Us!	109
8	The Masters and Masterpieces	121
	Seek and You Will Find	131
9	Unveiling Creation Design	135
	Change Blindness	136
	The Art of Distraction	138
	PC^2 – Politically Correct Practices / Personal Computer Abuse	141
	What Do You See Now?	144
10	Obstacles That Keep Us From God's Passion	147
	Unverified Information	147
	Leaving God Out	149
	Societal Misconceptions	152
	Weaknesses of the Peer Review Process	161
	About Our Thinking . . . and Barriers to Learning	162
11	Transformers	167
12	Hope in the Passion Creator	179
	The Other Side of Heaven	181
	A Passionately and Creatively Prepared Place	184
	A Passionate Prayer, A Passionate Praise	186
Closing		197
	The Renewing of Hearts and Minds	197
Addendum		201
Notes		205
Author Biography		213
Endorsements		215

Acknowledgments

This book would not have been possible without the faithful support of my wife, Tammy. She has taken on two part-time jobs to help see this endeavor through to its completion. She was able to take my out-of-the-box ideas and make them understandable to others.

A spirit-filled new-found friend, Alan Dyczewski, offered words of wisdom to help me keep the book on track. This was to be a mission for God, with God's Word as the foundation of everything I wrote. Alan's sensitivity to spiritual matters was invaluable.

Long-time friends, Tony and Sue Masevice, offered much needed encouragement mid-stream. Tony's in-depth knowledge of scripture and his gift of discernment were essential in keeping the project biblically sound. Sue helped address the expectations of the reader.

I would like to thank my daughter Lynnea. A a college student, she was a sounding board for readers in that age bracket. Lynnea also helped me to express this message in a way that brought the love of God to the forefront, rather than the judgment of God.

My son Jared is a leader in the making. He enjoys nature and discovers all that he can about the amazing creatures God has made. His energy and excitement is contagious and remind me to look at all creation with a sense of awe and wonder.

The prayer warrior on this project was Marianne Murray Dyczewski. I will always remember her faithfulness. The uncanny timing of outside influences that continually happened during the course of this project would astonish anyone. This was my heavenly encouragement to keep driving forward.

The wisdom and big-picture perspectives of Wayne and Marcy Sacchini, were of great help in the growth of my own spiritual sensitivity. I have learned to raise my spiritual antenna each time they hand me an article or something of interest because God using them to grow my faith.

Doug Rhode was kind enough to guide me in a few areas of research which gave the project more meaningful application. I am also grateful for Kevin O'Reilly for his kindness, wisdom and his ability to listen as he acted upon the desires of our Lord.

My editor, Linda Cizek, was Godsent. She patiently looked past my rough edges and saw the vision that God had put on my heart. Her interest in the arts and natural science, her creativity and her passion for life as God intended it to be, along with her God-given talents and abilities, fully blessed this mission beyond my expectations. I do not know where the project would be without her.

I am thankful for long conversations with Mike Murray and, although he is now present with the Lord, he will never be forgotten. As a patent contributor on cancer research, Mike unknowingly piqued my interest in science. This eventually led me to address some of the holes that are found in evolutionary teachings and the peer review process. His kindness and friendship made a deep and lasting impact on me.

I am extremely grateful for God's provision for our needs during the writing of this book. To all of those who answered God's call and provided encouragement, prayers and (sometimes even cash), I thank you. Without these committed supporters, I would have never been able to complete this work.

This book only became a reality by the generous giving of time and insight from all these individuals. My prayer is that God would continue to use these people of great influence to further kingdom building, while grace and mercy abounds.

Preface

When we experience the beauty of God's creation, it is natural to begin to contemplate how this world came into existence. It is in these moments that we are closest to unlocking the door to the room that contains God's incredibly creative attributes and artistry tools. And what a vast repertoire of gifts our Creator (with a capital C) encompasses! Some people have been granted the skill to paint, or to play music or to design things. But they require tools and inspiration. God has the ability to create something from absolutely nothing!
There were no bowls of fruit or beautiful seascapes or memorable characters in God's life to serve as His inspiration. When God Creates, He starts with nothing. In that sense, we're not really Creators of anything (with a capital C)! However, I would say that we are creators of lesser thingsWe take great pride in working with all that God has given us to work with and try diligently to do the very best we can. After all, He put synaptic receptors in our brains that fire eloquently with truly great thoughts! But none of these great creation thoughts can compare with the ideas of God, the Alpha and Omega. We are encouraged to "consider creation" throughout the Bible:

> *I remember the days of long ago; I meditate on all your works and consider what your hands have done.*
>
> Psalm 143: 5

Today, our days are overflowing with commitments and the pressures of everyday living. Unfortunately, a jam-packed and hectic life can burden our hearts and weigh us down, like an over-stuffed suitcase that we drag by our side as we trudge through a busy airport. Daily "busy business" can cause us to miss the blessings and insights that can come through by simply taking a little time to "consider these things." When we finally do, we can truly begin to see the Master's handiwork and creative genius. As a designer, I use man-made artistic tools that lead me through the design process. I am excited to share these with you so that you can engage in God's passion better each and every day! This type of relationship with God will fill your life in a new way and can bring a new sense of peace and joy. It's kind of like skipping; you just can't be sad when you skip across the yard. Try it sometime (the back yard may be better to keep the neighbors from talking)!

When we consider the care and utter brilliance that went into creation—the passion that God has for you and for me—it is amazing!

My initial hope for this book is that it will cause us to look at creation in a new light; that we might see creation as an event which points to God's complete plan for humanity. His Word directs us to study it. Recent scientific discoveries actually support it. And I pray that this book will be used as a tool, an actual agent of change, which will help you to discover the Creator's deep passion that was behind it all. That passion is an incredible gift, sent directly from the heart of God to you and me.

As we humbly consider the very beginning—the time before Genesis—our hearts may be touched in ways we may never have imagined. God was there, thinking and planning, before the existence of anything:

Father, I desire that they also, whom You have given Me, be with Me where I am, so that they may see My glory which You have given Me, for You loved Me before the foundation of the world.

John 17: 24

Before the mountains were born or you brought forth the earth and the world, from everlasting to everlasting, you are God.

Psalm 90: 2

When asked about our own ideas about how the world began, it is easy to default to a "just add water and stir" answer: God spoke and it happened. However, when we pause to really reflect on the planning that went on before God spoke the world into existence, the awesomeness of this act can fill us with wonder.

As we try work toward understanding the creative process, we can apply what we learn to the Biblical picture of God and His works. Then we can better understand His creation. We can recognize the incredible, loving forethought that God put into every divine act described in Genesis.

Furthermore, we will discover that God is still passionate about us. He calls out to us daily, in a very real and tangible way. He desires to show us He is close to us, as a father longs to be with his child. To live and die on this planet without giving credence to the God who created this universe will not only result in missed personal blessings, it will ultimately affect our eternal destiny—a tragic consequence indeed.

> *For since the creation of the world His invisible attributes, His eternal power and divine nature, have been clearly seen, being understood through what has been made, so that they are without excuse. For even though they knew God, they did not honor Him as God or give thanks, but they became futile in their speculations, and their foolish hearts were darkened.*
>
> Romans 1: 20–21

After God completed planning and designing the universe, creation commenced. By His Word, God spoke all into existence:

> *Then God said, "Let there be light," and there was light.*
>
> Genesis 1: 3

In the beginning, therefore, God was central to the abyss of the universe, less the stars, sun, moon, galaxies, and even space as we know it! The gospel of John opens with the words, "In the beginning was the Word, and the Word was with God, and the Word was God." He is eternal. God's other attributes include: omniscience, omnipresence and omnipotence! Our understanding is limited, but God has no limits. However, we might ask ourselves, is it possible to go beyond our current capabilities of understanding in our finite state? Can we push the limits of understanding as we investigate this topic, even when His creative power, genius and intelligence are inconceivable? If we can gain a stronger sense of God's identity and learn what His mission is, then perhaps the worship that we express back to God may be better focused and more passionate.

God reveals Himself in creation. Just by acknowledging the observable evidence in fossil, coal, and canyon formation in the recent volcanic eruption of Mt. St Helen's should make us pause and think. These findings support biblical creation. And we must be concerned that God's wrath is unleashed when men suppress truth:

> *For the wrath of God is revealed from heaven against all ungodliness and unrighteousness of men who suppress the truth in unrighteousness, because that which is known about God is evident within them; for God made it evident to them.*
>
> Romans 1: 18–19

Isaiah 2 reveals that it is in God's character to shake the earth in the day of our Lord. Is the shaking of the earth God's wrath being revealed? And if so, what should be our response? Again, we must pause and think this through.

Furthermore, the Bible teaches that God is a jealous God. In Deuteronomy, Moses writes,

> *For the LORD your God is a consuming fire, a jealous God.*
>
> Deut. 4: 2

God is jealous that the glory of His creative work is being usurped by those who purposefully (or inadvertently) advocate a "random chance" belief system, as in Darwin's Theory of Evolution.

"Glory is the supernatural signature of God." This profound statement from James MacDonald, senior pastor of Harvest Bible Chapel, stems from Psalm 19:1, "Heavens declare the glory of God." We should be the generation that gets out of the way of God's glory.

Solid science continues to reveal evidence that creates skepticism pertaining to Darwinian evolutionary theories. As of this writing, over 4,000 scientists and scholars across the globe[1] share this skepticism, along with another 1,000+ physicians and surgeons.[2] These numbers represent only those who have gone on record with their beliefs. There are also a notable number of professionals within the scientific and medical community that have chosen not to go on record or haven't been surveyed, who share in this skepticism.

The muzzling of new scientific discoveries which existing scientific models must also be addressed. Researchers must be able to reach logical conclusions unencumbered by the anchor of old-theory science.[3] This citation reference dates back to 1954. I believe it's time for a change.

I hope that when you are finished reading this book, you will understand that you are at the very epicenter of God's passion and that this realization would begin an ardent quest to learn more about our loving Creator. My second hope is that you will be able to look past the propaganda society is perpetuating and be able to simply ponder the wonder of creation in its purest sense.

Before the Beginning . . . God Designed weaves together many of life's unanswered questions regarding perplexing events. Finding a connection between the invisible and intangible spiritual realm and the visible, concrete world will help to settle and clarify many of these perplexities. I hope to demonstrate how art, science and the natural world all work together to reflect the glory and creative genius of the Ultimate Designer–the Lord of the Universe. I pray that this will allow you to develop a deeper relationship and appreciation for our Creator and in turn you will be able to see the world around you with fresh new eyes and renewed wonder!

You may even learn how to engage the creative and visualization process more quickly. I pray you will strive to be curious and to ask questions. You might just come away from this book seeing life from a completely different point of view!

The purity, the wholeness and completeness of heart, mind, body and soul that God intended for us before the beginning is truly something to contemplate. The old can become new again when we come to know God for who He really is: the Planner, the Designer and the magnificent Creator of all things!

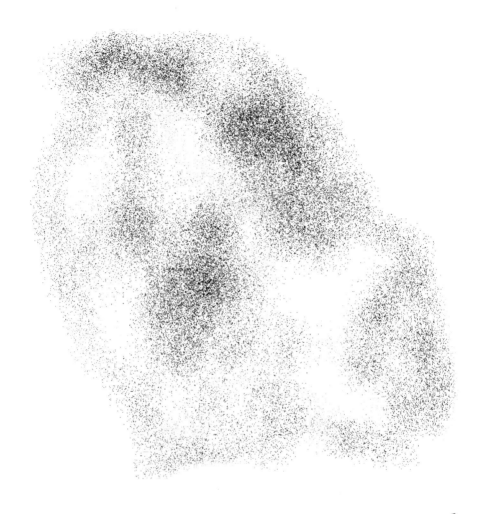

1
What Was He thinking?

What was on God's mind when He was creating? What prodded Him to create, conscious beings in His own image? What was God thinking about prior to Genesis, prior to all that we see—and don't see—underwater creatures, land animals, underwater plants, plants on the earth, plants that grow in the air, man, woman, the earth, the atmosphere, the planets, gravitational pull, the stars, the sun, the galaxies... everything! Why did He create you... and me? Why does He produce such heart-touching beauty everywhere? And what about the precision found in life-sustaining order? His awesomeness is in every crevice of existence! God is involved in all of creation.

> *For by him all things were created: things in heaven and on earth, visible and invisible, whether thrones or powers or rulers or authorities; all things were created by him and for him.*
>
> Colossians 1: 16

How deeply can we ponder this moment prior to tangible creation? Like a child who is about to look into a gigantic telescope with a special lens filter that allows him to peek into another dimension—and a timeless one at that—we are left breathless at the very thought of peering inside the mind of God.

But God gave us the ability to ask some very deep and intriguing questions. He has given us gifts of creativity and the ability to wonder. How far do we take His declaration that we are His true image bearers? And just how much of ourselves do we invest in our quest for truth? In this respect, we have been given a mission in the book of Matthew. When asked by the Pharisees which commandment of the Law was the greatest, Jesus responds,

> *You shall love the Lord your God with all your heart and with all your soul and with all your mind.*
>
> Matthew 22: 37

According to these words, we must give our all in striving to understand His creation and all things of God! He requests that we seek to know Him with every fiber of our being. Searching the deeper things of God (profound and bottomless) also takes a work of the Spirit of God, which may or may not reside in us.[1] *If* the Spirit resides then understanding will come more easily.

One of the first deep questions is "Why did He create us?"

Was God just bursting with excitement, wanting to share about His creations to come and divine understandings with someone else? In John 3:16, God is described as a compassionate God. Were we created so God would have someone to love ... an object of His love as stated in John 3: 16? God is beyond generous, sharing eternal life and all that accompanies this privilege. I don't mean to humanize God's feelings or expressions of His love, but God expressed intense passion for all of mankind by offering up His Son in a death sacrifice restore the fellowship that He once experienced with Adam and Eve in the Garden of Eden.

Why did He create us?

What was God thinking with such an intense act? The relationship is certainly one that we could identify with—a father and a son—but the sacrificing of that cherished child for the salvation of all mankind? This was an act beyond all human comprehension; a loving deed intended to save. Ultimately, this would become the single most impactful act of all time for all of humanity. I believe He also wanted to spark our curiosity.

Your eyes saw my unformed body. All the days ordained for me were written in your book before one of them came to be.

Psalm 139: 16

God was entering names in the Book of Life before He even formed the world! The verse above implies that there is a dimension and a beginning where time doesn't exist. That is such an amazing thought! In order to be able to do this, He had to give names to those before and after us. He knew we would struggle to comprehend this idea upon reading it. Could the verses be intentionally written for us to get a glimpse into this timeless dimension?

Congruently, we are always told we can't think in a realm without time. In the premier of the Star Trek TV series Deep Space 9, a time-independent alien discusses human existence as being "linear"—having a beginning, middle, and an end. The alien was observing the earth's time line, as though he wasn't part of it, like a third party observer. Richard A. Swenson M.D., is a physician and a futurist with a B.S. in physics, Phi Beta Kappa from Denison University. He states, "If God operates in more than one time dimension, it means that He can move around within our single time dimension and see everything happening simultaneously." Perhaps God sees not only our dimension that way, but all of creation, inclusive of time and space, the same way.

> God, in His awesome wisdom, even gave us the gift of the Bible—the unprecedented message of passion meant to lift the condition of our hearts.

Our challenge, as students of Bible passages and concepts, such as these, is to keep an open mind as we look into a place where all is still, and time has no meaning. If a clock were somehow placed in that dimension, the second hand would suddenly stop. There would be no hurry to do anything. However, a flurry of activity would surround our Creator as He is busy designing and orchestrating all life events; already knowing that it would all be recorded in the pages of His written record, from Genesis to Revelations, as well as each name written in the Book of Life.

But all of this was done with intentional and unfathomable love; with passion as it was meant to be. Pure creation! Pure intention! The Bible also reveals His absolute desire for true friends[2]... and thus, He created man. God wants us to be living in harmony with Him—in perfect orchestration as we follow His melody to fulfill His divine plan. Like the beautiful and unique instruments of an orchestra, of us has a different purpose to discover. Moreover, as we reach that point of discovery, we should seek God's will in determining what to do with that information.

He has given us independence and free will. We may want to consider why the awesome, incomprehensible mind of God would desire to create such independent beings like us; beings capable of deep thinking and creative action. He did this to confirm our individual identities. He knew it would take the sacrifice of time and energy to focus on the journey of putting Him first. Perhaps this could be a test of our genuineness in the pursuit of getting to know our Creator.

It took divine careful thoughts to create us and all that we see. God, in His awesome wisdom, even gave us the gift of the Bible—the unprecedented message of passion meant to lift the condition of our hearts. If we can look at things of creation in a different light and gain insight into the heart and intent of God, then we can humbly say that a new day has truly begun!

What was God thinking when He was creating? His thoughts are beyond what we are capable of grasping, He wrote his "love letter" in a form that would be understandable by humans. Just thinking about the fact that He allows us to read His inspired and holy thoughts is amazing. Words are almost useless in describing God's mind:

> *For My thoughts are not your thoughts, Nor are your ways My ways, declares the LORD. For as the heavens are higher than the earth, So are My ways higher than your ways, And My thoughts than your thoughts.*

Isaiah 55: 8–9

Who can fully grasp the infinite capacity of God's mind?

> *He has also set eternity in the hearts of men; yet they cannot fathom what God has done from beginning to end.*

Ecclesiastes 3: 11

However, God desires for us to utilize our finite and limited thinking capabilities to understand His infinite and unlimited ones. Only through our *humanity*—in its fullest sense—can His deity can be realized. And this is exactly how He intended it to be from before the start of creation.

Scripture tells us that God's thoughts about each of us outnumber the grains of sand on the earth. In his book entitled, *In the beginning was information,*[3] scientist Dr. Werner Gitt, former director and professor at the German Federal Institute of Physics and Technology, explains the incredibly intricate orchestration found in nature. Dr. Gitt comments on how we take for granted the presence of information that organizes every part of the human body, from hair color to the way our organs function. He makes it clear that life has an enormous amount of organized data that is processed and synthesized by 'an originator,' if you will. Another doctor, Richard A. Swenson, states in his book, *More than Meets the Eye,* ". . . if we observe design, it is not wrong to infer a designer."[4] Both doctors concur. Indeed there was (and still is) a process in place, but before it was set into motion, there was motive and very clear design.

> God desires for us to utilize our finite and limited thinking capabilities to understand His infinite and unlimited ones.

> "... if we observe design, it is not wrong to infer a designer."
> Richard A. Swenson, *More Than Meets The Eye*

How painstakingly meticulous the designer of the universe was in His work! God went through the trouble of personally naming all the stars and He even knows if one is missing. Moreover, the power of His omnipotence is incredible. For example, He asks us to pray to Him. There are roughly seven billion people on the planet, each with a potentially long list of prayers. Yet, we have a God who can handle the volume. Not only do we have a God whose thoughts are far beyond our understanding, but also one who has a capacity to store, process, sort, analyze, create, trouble shoot, plan and initiate a series of events for a common good. He coordinates these events while simultaneously preparing hearts to learn and grow closer to Him as He lovingly guides us through this earthly life! Perhaps the most amazing element of it all is that each of our human prayers intersects with His pre-determined will, inclusive of His ability to move or inspire others, through His patient teaching, with His gentle nurturing, and in His perfect timing.

> "There are bigger things going on here than you or me."
> Peter Parker, a character in the movie, *Spiderman II*

A perfect illustration of this idea was developed in the movie *Spiderman II*. To seek revenge for his father's death, Harry (son of the enemy) confronts Spiderman, (whom he now knows is Peter Parker). In response to Harry's accusation, Peter replies, "There are bigger things going on here than you or me." This salient thought should be considered continually as we decipher God's plans for our existence and attempt to comprehend things beyond the daily grind of life; a whole other level—a bigger picture—exists.

So what are some of the 'bigger' things that have boggled the human brain for centuries? The visibles of life include the origin of life itself. How does one create something—anything—from absolutely nothing? Now what happens if it's invisible? Some of the invisibles of life include: gravity; electromagnetic force; the strong and weak nuclear forces; consciousness and instinct. All are needed for life! Just because we can't fully grasp these "bigger things" doesn't mean we should ignore and not address them.

In regards to the spiritual dimension, we wonder if there is a Heaven and a Hell. Is there spiritual activity going on around us? Is there a divine life-plan in place that is reliable?

We have air, water and food which all "magically" provide sustenance and are necessary for life! We were even given senses to experience this incredibly complex world and communicate ideas with others. God made a sun for light, energy, regeneration and plant growth. This is all proof that God wanted man to exist and thought of these things *prior* to the creation of Adam and Eve. Everything around us sustains our earthy bodies! God had a great *desire* for our company, but where did that come from? God is self sufficient! Why did He desire us to be in His presence, sharing the cosmos and definitely making His existence much more complicated? These are the questions that arise from the mysteries of creation. And they are pertinent questions to cause wonder and challenge us. God didn't give these answers directly for a reason. It seems like a huge, unfinished puzzle has been designed and created, and we need to find the missing pieces.

> How does one create something—anything—from absolutely nothing?

One of the themes woven through the Bible is the notion of fellowship with God. He surely doesn't require our company; rather it is we who need Him. The relationship desired by God must be genuine. After all, who wants a friend that hasn't made a personal *choice* to be your friend? James 4: 4 clearly states that there are two sides; we can choose to be a friend of God or to be His enemy. There is no middle ground. It is not unfair for God to want us to be on one side or the

> Since God is eternal, it stands to reason that He would not want the friendship to end upon our body's expiration.

> The devil's goal is to distract and deceive.

other because He gave us ... everything! Since God is eternal, it stands to reason that <u>He would not want the friendship to end upon our body's expiration.</u> That would be a very brief and shallow relationship. His blueprint included an endless bond for eternity, inclusive of an everlasting body and soul:

> *So will it be with the resurrection of the dead. The body that is sown is perishable, it is raised imperishable; it is sown in dishonor, it is raised in glory; it is sown in weakness, it is raised in power; it is sown a natural body, it is raised a spiritual body. If there is a natural body there is also a spiritual body.*
>
> I Cor. 15: 42–44

We are of the utmost importance to our Creator and He loves us completely. Along with the intentionally-designed, unfinished puzzle, He allows distractions, trials and obstacles to test our faithfulness, to develop our character or to humble us. When we discover why these events occured, another puzzle piece is uncovered.

To understand the trial puzzle-piece the Apostle Peter encouraged other Christians to be strong as the persecution of Christians spread throughout Asia:

> *These trials will show that your faith is genuine. It is being tested as fire tests and purifies gold—though your faith is far more precious than mere gold. So when your faith remains strong through many trials, it will bring you much praise and glory and honor on the day when Jesus Christ is revealed to the whole world.*
>
> I Peter 1: 7

The bible is filled with many stories of people who suffered tribulations, such as Job, Moses, David, Jonah and even God's own son! Trials can change our perspective and we will probably get to know God better working through these challenges.

These numerous distractions can be dangerous and are orchestrated by another conductor—of sorts. God has allowed a fallen creation –the devil- to emerge and to assert tests that may have eternal ramifications. The devil's goal is to distract and deceive. And he does this cunningly. Remember, the devil doesn't want us to solve our puzzle because with each placed piece, we are closer to seeing the completed picture.

Jesus explains that the enemy who sowed weeds in the farmer's field is a simile for the *devil* and his deeds. When we develop a correct perspective regarding how God uses our troubles, we can learn to practice being faithful to God and ourselves as His people. The deceiver's work is invisible to our physical eyes but recognizable (though not always immediately, like a weed's root system underground), through the discernment of the Holy Spirit within believers. The devil has the power to soften our resolve and even work against the unbeliever.

The devil's role includes a mission to distort our uniqueness. For example, our appearances in skin or hair color or shape and size differences can be so different that these can scare or intimidate. We often can prejudge a person based on appearances, our own previous experiences or disinformation we may have received. Satan is referred to as "the father of lies" in the book of John. Jesus' body was as He intended it to be. If this is any indication of our heavenly bodies, believers have something to look forward too. Our bodies will also last for eternity, and will be given freely and lovingly by God, our father.

Many of us will be given eighty or more years on this planet, which is plenty of time to exercise our ability to make choices. Perhaps the most important decision we can make is to spend some of those precious days thinking about the things of God.

In fact, some have already decided to reject the things of God and God knows it! The Word of God states that the things of God are "foolishness to those that are perishing." These are difficult words to hear, but they are pertinent for understanding "before the beginning" planning. God is the ultimate playwright, if you will. He has put into motion a spellbinding drama of good vs. evil and the title of this very significant play is life. He was already aware of those who would turn their backs on him before He even held the first casting call. He knew who would play the roles of the murderers, the thieves, the liars, and those whose lives would be consumed with hatred for his fellow man.

A question that plagues many is: How could a loving God allow such evil to happen in our world today?[5] Just after the beginning, in the Garden of Eden, everything was as God intended, existing in perfect peace and harmony. Sadly, we disobeyed God and chose evil, opening the porthole for the devil to slither into our world. We are all born with a sinful nature. If we don't accept that salient element of our personalities, ramifications will emerge. Sin can be very destructive and it separates us from God. It's very hard for most to see it as God sees it.

> Sin blocks us from possessing the ability to discern truth and, worst of all, it prevents us from getting to know God better.

But your iniquities have made a separation between you and your God, And your sins have hidden His face from you so that He does not hear.

Isaiah 59: 2

I look at sin as a wall, or a blockade. These sins interfere with our ability to have an abundant life. Sin blocks us from possessing the ability to discern truth and, worst of all, it prevents us from getting to know God better. Some of the names we've given to these blockades include arrogance, pride and idolatry. These blockades act like an undetectable veil to eternal truths. They allow lies of every sort to distort the very make-up of who we are. The good news is that these blockades can be discovered and exposed in our lives as we identify them. Once identified, we need to seek ways to overcome these hurdles through prayer and reading God's Word. This passage in Matthew is of great help during this process:

> *But seek first His kingdom and His righteousness, and all these things will be added to you.*
>
> Mat. 6: 33

This intentional right-thinking on our part can launch blockade-busting missles to bring about real change.

> *But we all, with unveiled face, beholding as in a mirror the glory of the Lord, are being transformed into the same image from glory to glory, just as from the Lord, the Spirit.*
>
> II Corinthians 3: 18

I mentioned earlier that God is the divine playwright. Let's jump to the end of the play for a moment. After physical life has ended and all of life's distractions (blockades) are set aside, we will be in eternity with God. Understanding the very nature of God, being transcendent (set apart from all that we know) in Holiness and all that is good. We will find ourselves in the presence of a loving Heavenly Father who ultimately cares very deeply about us. He cares about the condition of our hearts now, in the journey, and later in His presence.

Awesome Minds

Because we are created in the image of God, we have a mind similar (in some respects) to His. I can't help but to enter a state of awe when I think about the mind of God. He had the incredible mental power to create a human brain, with all of its marvelous complexities. How did He do this? What a design challenge! The human brain is a highly technical and complex organ. It can differentiate things, it can manage an entire central nervous system and respond to all stimuli through sensory nerves for sight, sound, touch, speech, hearing, smell and my favorite . . . taste![6] With the brain, we are able to read, interpret and store information that is gathered from each of the nerve's senses. To be able to play back and transmit these stored memories/instructions to the appropriate body part is beyond amazing. Our brain gives instructions to our appendages. It runs our organs. The brain tells micro bodies when and where there is an infection. Then sends an army of antibodies to the battleground to fight in the name of our health!

According to Dr. Swenson, within a single gram of brain tissue, "there are 400 billion synaptic junctions"[7] that conduct healing and bodily functions above and below our skin. These junctions provide a warning when something is too hot or too cold with our "450 sensory cells on each square inch of our skin."[8] This brain tissue, "with just under, 200,000 miles of neurons and dendritic connections,"[9] has the ability to link information, which we call knowledge. Our minds allow us to be able to use that knowledge to create, teach, protect, feed, inform, lead, fight or flee. When we use our knowledge correctly, we call that wisdom. This gray matter with a "storage capacity equal to that contained in 25 million books"[10] also has the ability to discern, salivate, procrastinate, pursue, get happy, feel a dozen emotions and trigger a tear that communicates sadness to others. It can pick up multiple communication signals from others not just through words alone but also through body language. A wink or a rolling of our eyes, a simple smirk, the raising of a single eyebrow, a crossing of arms, a lofty glance or a shoulder shrug can all be subtle ways to communicate. It's designed to be drawn to interests, drawn to find a mate, procreate, speculate, animate, formulate, differentiate, articulate, agitate, anticipate, appreciate or celebrate. And each of our brains is designed to be unique from all of the other brains on the planet!

Our brain is designed with the ingenuity to create and invent on its own, to truly think independently, to problem solve, to fix, to build, to discover and to desire and seek adventure. We have the ability to conquer our fears and to take on different levels of risk. Our mind desires and enjoys a good story. It has the ability to create a story, and desires to know the conclusion.

The brain has to be housed in a safe place to protect us from everyday knocks, bumps and bruises. Hence, we have a skull. The brain tells us when to open and close our eyes and our mouth. It tells us when we are hungry and thirsty. It will break even break down if not properly loved. Could there be a message to us here? Again, we are brought back to considering the bigger things that are tied to our ultimate friendship with God.

In fact, the Bible teaches that we have an emptiness built inside that can only be filled by God. We are truly mysteriously and wonderfully made! And we have been created to love and be loved. As Augustine once said, "God has made us for Himself, and our hearts are restless until we find our rest in Him." In the 43rd chapter of Isaiah, God claims His people:

> *Everyone who is called by My name, And whom I have created for My glory, Whom I have formed, even whom I have made.*
>
> Is: 43: 7

Loving Us Through Design

There is a powerful God-given trait that we all possess which is a vital attribute in reaching our fullest potential. For some, discovering it can drive them to levels of accomplishments they never believed they could achieve. But for others, it can be a dangerous, all consuming monster that devours life in one gulp. Two men in the Bible who exemplify these extremes are Paul and David. Both possessed this trait in abundance, but at times their hearts were ruled by something as far from God as one can get. What is this God-given gift that can raise us up to new levels of understanding one day and, if we're not careful, cast us into the valley of destruction the next? The answer is . . . passion.

Passion

Many would agree that our God is a God of passion. Webster defines passion as "an extreme, compelling emotion; an intense emotional drive or excitement; a great anger, rage, fury, enthusiasm, strong love or affection." Another synonym for passion is the word zeal. According to BibleGateway.com, this word is in use anywhere from twice to 36 times, depending on the translation (King James, The Message, NIV, Wycliffe, etc). However, the word zeal is seldom used in modern language because it's an older word that is not often utilized in day-to-day conversations. Passion, like the word zeal, carries an intensity and depth in its meaning.

Other descriptive terms for the word passion include: "a strong, fervent desire, an ardent affection, enthusiastic devotion to a cause, ideal, or goal and tireless diligence in its furtherance."[11] There cannot be passion without emotion. When God said that He wished that "none would perish" there was great conviction and anguish behind those words! When the Lord's people were praising Jesus as He entered Jerusalem as a king, the Pharisees demanded that His disciples be quiet. Jesus replied that if the disciples kept silent, even the rocks would "cry out." One assumes that a "cry" contains great emotion, right? But how can rocks have emotion? The Lord personified an inanimate and seemingly useless object and gave it a human capability to make a point.

> There is a powerful God-given trait that we all possess which is a vital attribute in reaching our fullest potential.

The cartoonist Charles Shultz allowed his character Charlie Brown to receive a "lump of coal" in his Christmas stocking. This became the ultimate symbol of a worthless gift! God uses that which is useless. He lifts up that which is the lowest. If lowly, useless rocks would cry out to praise the Lord, what would other creations of God express? What would the majestic mountains and explosive oceans be saying if they could speak? What utterances of the glorious Creator would the trees and hills whisper? Whatever they would say, I believe their voices would be filled with *passion*!

> What would the majestic mountains and explosive oceans be saying if they could speak?

Because *we* are made in the image of God, He made our minds, hearts and souls *passion capable*. What an awesome trait to be hard-wired into our very core! No other trait permeates our being as passion does. And when God is behind the passion (in its truest sense), man can *move mountains*! When we are impassioned, obstacles and excuses seem to reduce in size. We find ways to make big things happen. When we discover who God intended us to be—when we've discovered our purpose—we become focused and driven! Depressive tendencies can vanish; excitement and ultimately, real contentment can replace it. In the book of John, Jesus said, "I am come that they might have life, and have it abundantly." But sadly, although we all possess this great gift, many of us inadvertently suppress it, hide it, or maybe even experience shame from it. I am afraid to be passionate at times, for fear of having my hopes and dreams squelched or dampened. Fear often comes from our inner self. It is not of God. Fear is an ugly thing and the fear of failure can rule our lives and prevent us from reaching our divine purpose.

> He made our minds, hearts and souls *passion capable*.

Most of us have probably never thought about the *power* of passion much, or the role that the Holy Spirit plays as we try to navigate the road of life. However, if left unchecked, passion can grow into something that is too wild to harness. In fact, it can even become an idol if we are not careful. But under the control of the Spirit, great things can be accomplished, once our sinful nature is overcome. If we are willing to let go of other distractions, blessings are sure to follow. Seeking the Lord *passionately* helps us to discover all that God has in store for us.

In his book entitled *Primal*, Mark Batterson talks about the raw intensity needed in our approach to God—with primal compassion, primal wonder, primal curiosity and primal energy.[12] Batterson further goes on to say that souls are lost because of the lack of passion or enthusiasm, even if it is a believer's soul.

David was a man after God's own heart, but David had a dark side. He was capable of first-degree murder. He also had lust in his heart that drove him to commit adulterous behavior with Bathsheba. David plotted well in advance to have Bathsheba's husband killed, so his adulterous affair would be concealed. However, David also had a thoughtful and curious side. By reading the words that a remorseful David later recorded in the pages of the Bible, we can see that he rather easily expressed his feelings, emotions, attitudes and interests.

> When we are impassioned, obstacles and excuses seem to reduce in size.

David took the time to repent. He took the time to praise, worship and think deeply about his God. It is thought that David wrote 73 songs of praise in the book of Psalms,[13] many of which have inspired worship songs which are still being sung in churches today—over 3,000 years later!

Paul was a murderer of scores of Christians, and it seems as though he was quite glad to do it. However, one day God struck him with lightening and took his sight away. Finally, after Paul considered the things that God had said to Him, Paul's sight was returned to him and he was transformed. Later, Paul led many people to God. He was filled with such conviction that he was even imprisoned for his beliefs! He authored several books that are now part of the New Testament which continue to inspire many believers with his incredible witness and devotion.

Why would God overlook such vile behavior and still use mankind in great ways? Through these stories, we see how God loves us even when we are buried under mountains of character flaws. He sees our potential! God is showing us that He can use anyone He chooses; even you and me!

> **Why would God overlook such vile behavior and still use mankind in great ways?**

The Lord is compassionate and gracious,
slow to anger, abounding in love.
⁹ He will not always accuse,
nor will he harbor his anger forever;
¹⁰ he does not treat us as our sins deserve
or repay us according to our iniquities.
¹¹ For as high as the heavens are above the earth,
so great is his love for those who fear him;
¹² as far as the east is from the west,
so far has he removed our transgressions from us.
¹³ As a father has compassion on his children,
so the Lord has compassion on those who fear him;
¹⁴ for he knows how we are formed,
he remembers that we are dust.

Psalm 103: 8–14

> **When God said to love Him with all of our heart, mind and strength, He expects us do it passionately.**

Man was granted the gift of passion by God, but man is obviously influenced by self and the enemy. David passionately lusted after Bathsheba and committed murder. Paul, a Pharisee under Jewish law, was zealous enough to persecute the early church, even to the point of taking lives. Through the working of the Holy Spirit, Paul and David were transformed and lived their lives *passionately* in a new way. They lived life to the fullest, according to God's will, rather than their own. God saw this and was pleased.[14] He set David and Paul apart and never left each man's side.

An example of a modern day man of passion is the late Steve Irwin, crocodile hunter and wildlife warrior. An interview with Steve's wife revealed that when Steve was with animals, nothing else mattered! He was focused, in love and steeped in mission. For Irwin, crocodiles represented dinosaurs that were not extinct but left for us to discover, study and wonder about. Steve wanted to preserve and protect crocs, as well as many other animals.[15]

Irwin's family didn't sweat the small stuff. It was in their genes to get dirty, get wet and get engaged. The other items didn't matter. Animals can be dirty, wet and very exciting! When we examine the design of the myriad of animals that exist today, it can take our worship to a deeper level, because we begin to think about the fact that there is a loving and caring God behind it all.

> Paul and David were transformed and lived their lives *passionately* in a new way.

I couldn't find any information that indicated anything about Steve's spirituality. I don't know where he stood with God. God knows and it's ultimately God's business. Hopefully, Steve knew there was a perfect designer behind this wild world before he lost his life while exploring the ocean. Steve's example of engagement with creation remains an inspiration to me.

God designed us. He can use us, even though we are weak and far from perfect. He intentionally designed us this way in order to accomplish His will. He can love us through our unique design. We are a part of His story. We are called to live life *passionately*! When God said to love Him with all of our heart, mind and strength, He expects us to do it *passionately*.

Dr. Jobe Martin is another modern day example of a passionate professional. Dr. Martin has written many resources detailing the handiwork of God in the creation of animals. He is interested in communicating the love of God exhibited through the design of God's creation. Dr. Martin brings to light animals with very unique design characteristics. He pushes us beyond traditional thinking about such things.

Dr. Martin disagrees with the concept of macroevolution, which teaches that animals evolved from one specie to another specie. Dr. Martin explaines how that certain animals disprove the theory through their design. These animals must have been complete in design and possessed a finished form from the very beginning, or they would have died before they had a chance to evolve!

Let's look at one of these animals Dr. Martin mentions in his Incredible Creatures That Defy Evolution DVD. IIn the case of the giraffe, Martin points out that it would "take a great pressurized pump to

pump blood over six feet straight up through the giraffe's neck to get to the giraffe's brain." The same pressure necessary to pump the blood up seven feet to the brain would explode it when the giraffe bent down to drink. But God designed a solution for that.

The giraffe's neck is so awesome in design that it has six valves in it! Each valve closes consecutively as the giraffe lowers his head. The design isn't yet complete, for the giraffe wouldn't be able to stay down long enough to finish a drink of water before passing out! So the giraffe has a "sponge" at the end of his spine that compensates for the shutoff valves by supplying the needed blood. This balances his equilibrium, so he is able to lift his head very rapidly without passing out and make a quick getaway from predators. Dr. Martin uses this example, as well as many others, to show the necessity of a complete design for animals to survive.[16] If this were the case, a half developed or half evolved giraffe could not survive its first drink of water and could not reproduce. An underdeveloped valve system in the giraffe's neck would explode the brain. Also there is no fossil record verifying these transitions and to verify macroevolution. The correct design had to be in place from the first giraffe.

> To create awe is a practical thing for God to do.

Looking closely at the design of the giraffe, we see a passionate designer. God's love is expressed here. The giraffe is beautiful in design and color. It has a very unusual shape. Each of the giraffe's spots are unique, just like our fingerprints. It's stomach has four chambers to digest the lofty vegetation. It is the tallest animal on the planet and was made intentionally to reach great heights. It has the long legs to go with the long neck. It even has furry horns atop its lofty perch, perhaps only to make its face look both silly and friendly. Children giggle in amazement at these creatures! What practical purpose does this giraffe serve except to amaze us? Perhaps God did this so we could grasp the magnificence of His handiwork in our human minds. Why would God do this? The answer is simple: To create awe is a practical thing for God to do. As part of His unfathomable plan to establish a relationship with us. He proclaims Himself the supreme and sovereign God who set everything—*every living thing*—into motion, before time began.

There are many animals that challenge man intellectually, some directly and some indirectly. Dr. Martin covers several more in his material. When we focus on the animals with characteristics that are inherently designed with a message, lesson, example or a practical observation, we are blessed with new insight. It becomes strongly apparent that a great 'dad-like' God, a 'dad-like' designer—is behind all that we see! Recognition of this intamacy brings glory to God, and this is a good part of what it's all about! Chapter seven highlights seven more animals that are very complex in design, yet simply amazing.

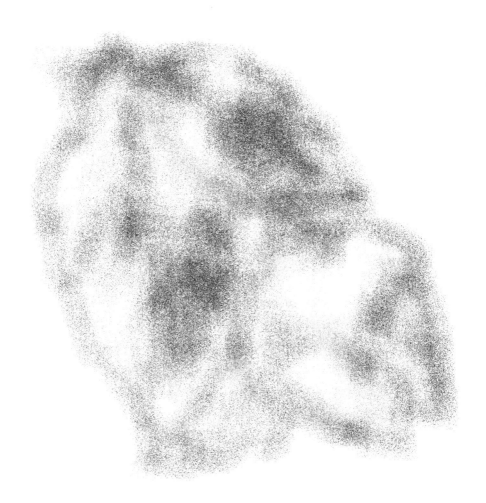

2
God's Heart in Communication With Man's Heart

Michelangelo's painting of God's hand reaching to touch man's hand is quite astonishing, especially once you learn about one other element of the painting. In the background, behind God, there is a slightly camouflaged shape of a brain! According to the guide at the Sistene Chapel, this is believed to represent the mind of God. Michelangelo's talent (like that of Beethoven, Bach and others) was so unique and beautiful that it has been difficult to fully credit him alone as the beginning and end of his work. Their work has been widely recognized as divinely inspired.

24   Before the Beginning

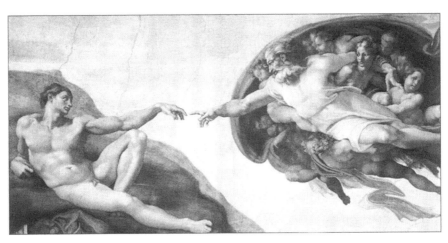

Michelangelo's, *The Creation of Adam*, fresco painting.

When placing the subject of creation of human beings under the lens of the microscope, we find ourselves asking another question: What *inspired* God to create us? We are his children and He longs for us. Those of us who are parents know the passion we feel toward our children. We long for them. We enjoy the wonderful feeling of holding our children in our arms. They ride perfectly on our forearms as newborns and a toddler sits neatly on the mother's hip; the fit is perfect!

The bond between parents and children is so precious. God desires this same intimate bond with us! When a parent sees their own characteristics (such as dad's sense of humor or mom's compassion towards others) in their own child, the parent-child connection is good for the soul. I believe that when God sees the spirit of His own attributes reflected in His children, He is certainly pleased. Scripture addressed the depth of God's longing for this in the book of James:

> *He jealously desires the Spirit which He has made to dwell in us.*
>
> James 4: 5

Can you imagine having this as a design challenge? In His desire to make a brain capable of feeling connected, God met and surpassed His goal! Wow! Our brains are designed to "feel connected" to our loving Father and that is such a wonderful thing!

As I mentioned earlier, man was designed with feelings and emotions as a central core of our being. Some argue that our emotions are what separate us from the animal kingdom; that they are what make us "human." But many animal lovers will disagree with that line of argument. We can probably agree on a few things about this aspect of who we are: Feelings are necessary in life. They can be fickle and mislead us at times. And when a coworker or spouse speaks to us about certain things, emotions can be difficult to read. But just like sorting out the feelings of others through picking up on things like facial expressions, body language and voice tone, there are also clues (ways of learning to "read") God's heart.

> The bond between parents and children is so precious. God desires this same intimate bond with us!

We see hundreds of references to man's heart in the Bible. Yet, it is somewhat amazing that there are minimal direct references to God's heart. Only 26 of 598 references to heart in the Old Testament allude to God's heart.[1] However, we are continually encouraged to *know* God. The book of Psalms advises that we come to know Him through His law, commands, statutes, precepts and decrees. Many of these refer to our approach to God and we may utilize these areas to build a foundation of divine understanding. Through studying the Scriptures, we can begin to comprehend the heart of God.

> In His desire to make a brain capable of feeling connected, God met and surpassed His goal!

The Lion Witch and the Wardrobe is a movie which was based on the popular series of books that were written by the Christian author, C. S. Lewis. In the story, Mr. Tumnus, a mythical fawn, is befriended by Lucy (one of the young protagonists). Mr. Tumnus tries to explain to Lucy the comings and goings of Aslan the Lion (a symbolic Christ figure in the story). He tells her, "After all He is not a tame lion." It is difficult for Lucy to grasp this very new world with such a seemingly aloof king, who is so cherished. Similarly, the Psalmist states, "In all your ways acknowledge Him, lean not on your own understanding." Though it may be a challenge to love Him who we do not at times understand, this is what we are called to do. Philippians 3:10 reads, "That I may *know* Him, and the power of His resurrection . . ." These verses invite us to intimately know Him with all of our being and to rely on God and His Word, rather than our own understanding. This is where the amazing faith-walk begins. By leaning on His promise, and by letting go of our human understanding, we can learn to fully and wholeheartedly believe that God loves us completely. Through the stories in the Bible, we see His compassion, mercy, loving kindness, and the ultimate gift and sacrifice of His Son, Jesus. This is the "unspoken" message to our hearts, that the heart of God is very good, indeed.

Let me understand the teachings of your precepts;
then I will meditate on your wonders.

Psalm 119: 27

The Message

In 1990 during a heart-wrenching trial of survival in my graphics career, my Pastor came to me and said, "Jim, I do not know what this means and I have never had this happen before, but in my sleep, I believe the Holy Spirit told me to tell you that He cares more for the condition of your heart than the condition of your business." For two years, my workload slid down to the point where I was no longer able to support my family. Although this happened over 20 years ago, I have never forgotten his words. It didn't have the meaning then as it does today. Through our intense trials, and in times of reflection, I have been overwhelmed by questions that refer to the "condition of my heart." Did I lose perspective because I substituted the value of my business over of the value of my heart? Yes indeed!

How is your heart? Would God say it was "in shape" or could it use a ... little "work out"?

King Solomon, considered by many to be the wisest man on earth, gives us a stern warning in Proverbs. He says, "Above all else, guard your heart, for it is the wellspring of life." Protecting our heart ranks high with God. Why did King Solomon say this? Our Designer God has made us vulnerable to sin. A state of innocence was given to us before the fall of Adam and Eve. We are designed to make the choices which show our genuineness in our friendship to God. Also, it's good to learn from these lessons and tests. As we do, it can make it easier to grow spiritually as a child of God and more of life's puzzle pieces begin to fit together.

During our time of trial, we were dealt a crushing blow financially. In the field of communications, when signs indicate the economy is declining, industries cut advertising dollars. When the economy improves, advertising is usually the last to be restored. Contacts in this field can be quite transient, you have to keep reselling yourself. And, to add to it all is ever-changing software! I remember the local paper quoting me in their quote of the month. I stated that the economic slowdown was caused by everyone trying to learn their new software program, not a reduction in factory orders, the usual reason for a turndown in the 80's and 90's. In these situations, it's easy to lose sight of the fact that I am a child-of-the-King. This can be very devastating to believers and non-believers alike. People have even taken their lives when they feel they have no way out. I know what it's like to have my back against the wall and live with the frustration of not being able to pay bills on time and the thought of bankruptcy looming overhead. It's a very tough place to be! But today, I rejoice in the fact that I used the tools described in the Bible to survive, to keep myself literally alive! One of the verses that I hung onto then (and now) is . . .

> *For I am confident of this very thing, that He who began a good work in you will perfect it until the day of Christ Jesus.*
>
> Philippians 1: 6

This verse reminds me of the intimate involvement of a loving father. Having lost the presence of my earthly father through divorce at an early age, remembering the Creator's great love for me has made a tremendous impact in my life today.

God says that there is vulnerability, a weakness, and a tendency to drift away from His special and divine plan for us. Philippians 4:6, another survival tool, tells us not to worry about anything, but to bring our prayers before the Lord. And if we do so, the "peace of God will guard your hearts and minds in Christ Jesus." Some assume this "peace" to mean the Holy Spirit, but what a delightful message we are given here! The passage indeed infers that there is a desire or even passion for God to communicate to our hearts to safe-keep

> **God originally wired us to be *heart sensitive* and to, in essence, to be our brother's keeper.**

His creation. Do we need safekeeping? Absolutely! When our hearts are broken, we feel it. In fact, if we are not grounded in a relationship with the Lord, our life course may be completely derailed. Sometimes, we are able to muster up the willpower to get back on track and other times, God's sends someone to help us out. I am learning to manage these healthy concerns.

The *Andy Griffith TV Show* was popular in the 1960s and 1970s. (A heart felt thank you goes out to Andy and the producer for their work. Andy is now present with the Lord). Viewers were swept away by the nostalgia of the small town of Mayberry, North Carolina and all of the memorable characters that lived and worked there. Barney Fife, Otis, Ernest T. Bass, Goober, Aunt Bee and Opie were all colorful characters, but it was the Sherriff Andy Taylor who always had a kind word (usually based on a biblical principal) to steer folks back to the right path. Everyone looked to the Sherriff to rescue them from daily dilemmas and to make sure everyone was safe and sound. Andy had a way of caring for others in a warm and tender manner that no one else possessed. This was truly the "feel good" show of the era. Sometimes, it may seem as though these characteristics have all but disappeared in this world of climbing corporate ladders and keeping up with the Jones.' But God originally wired us to be *heart sensitive* and to, in essence, be our brother's keeper.

As God was developing the story of Life (with a capitol L), He was careful to involve all facets of great drama: a great story line, conflict and resolution, and internal and external conflict. In our world, there is good and evil, beauty and ugliness, truth and deception. The saga of Life has depth and breadth, and it is filled with mystery and intrigue. But, as in all stories, there is a message.

In fact, God doesn't have just one message, but an eternity of messages. Some are big picture items that involve

> If a thought about God warms your heart, comforts you, directs you, convicts you, teaches you, or makes you question or dig deeper, God is pleased. His motive is to lead us to Him.

> In the concept of communication, there are only three elements: a sender, a receiver and a message.

our destiny, step-by-step, through choices that we make. Some messages are small, but still significant. If a thought with God in mind warms your heart, comforts you, directs you, convicts you, teaches you, or makes you question or dig deeper, God is pleased. After all, His design gave us inquisitive minds. His motive is to lead us to Him.

> *After you have suffered for a little while, the God of all grace, who called you to His eternal glory in Christ, will Himself perfect, confirm, strengthen and establish you.*
>
> I Peter 5: 10

In the concept of communication, there are only three elements: a sender, a receiver and a message. God invites us to speak to Him through prayer, and He craves one-on-one time "without ceasing," according to chapter 5 of Thessalonians. Yet, when it comes to communicating with God, many fail. We are responsible for a break down in the communication process, because we stubbornly refuse to access the power that can come through prayer; we miss out on guidance, strength, healing and hope. He makes it very easy. He just simply says to pray and our loving Father will hear our songs of rejoicing and cries of remorse. He is patiently waiting to hear from us.

> *But know that the LORD has set apart the godly man for Himself; The LORD hears when I call to Him.*
>
> Psalm 4: 3

Most importantly, what does our heart need to hear—or perhaps experience—when we cross the plane from the physical world into the spiritual world via this amazing communication line?

When my son was five, he would come to my wife and me for a morning hug. He did this with no encumbrances; he just purely desired to feel close to us. This loving act was placed in my son for me to cherish. Even today, I treasure this memory.

When I was a toddler, I remember climbing in bed just wanting to be close to my parents and my dad getting very angry because of this. I remember leaving and being a little heart-broken. Today, I really don't blame him, because there were five of us kids, and he needed his space. However, it did hurt me to be unable to seek their comfort in this way.

So when my son was a toddler, each morning that my son came in to give us a morning hug, I welcomed him. The healing message of acceptance and comfort that it brought will not be forgotten! This simple act of my son coming to me for affection, and my accepting him is a great illustration of the way God desires us to work within the communication model: We enter His room and send the message that we need to feel loved. He receives that message and responds by drawing us closer to Him. The "message" is received from our heart to His and He responds back with love. It's that simple.

Another example of how God ministered to my heart was during another economic downturn. As an artist, being somewhat sensitive to things, I can't help but to continue to examine myself. If you are a creative person, you may understand that sensitivity is wired into our DNA. It can be a gift but also can be a curse. Door after door were being closed in the graphics field. The battle of self-worth begins once again! In an effort to seek work the resume was redesigned and sent to the multitudes, and it was met with very few replies. Was it the economy? Have my abilities become outdated? Round and round I would go! The pain of the never-lifting questions would plague my brain day and night! The mental anguish and frustration wouldn't go away! The house was being foreclosed on. Death by a thousand cuts revisited! I finally stopped and asked God what I was missing. It then occurred to me that I did not fulfill the plan that I had agreed with God on for our new house. I was to use the house for *His* work. I hurried back on the *Before Creation* manuscript. Finally, I understood: I wasn't giving my *full attention* to His call. To my amazement and shock, the chapter outline that I started five years earlier had included what we had just spent four years doing in farming! Two specific titles were "The Beasts About" and "The Garden Around Us." We had no idea how we would utilize our property to continue the "agricultural use" requirement for property tax reduction. I believe that God wanted to show us specific details of creation design that Jesus used in His parables and teachings so I could write these chapters with similar care and intimacy in which He created!

> **Finally, I understood: I wasn't giving my *full attention* to His call.**

My heart has been worn out from all the uncertainty that I felt was raining down on my life. God knew this. My wife went on a timely weekend trip to Washington D.C. with a dear friend. It turned out that our friend knew the creation researcher who was a national broadcaster for Moody Bible Institute. I realized that nothing happens by coincidence! I asked the researcher if he would consider taking a look at the book that I wrote which was certainly inspired by his work. We had a common desire: to bring to light God's genius as both Creator *and* Designer. He responded by writing a very warm review for my book. God knew that *this* would minister to me and lift my heart. I will be eternally grateful for God's caring for the condition of my heart!

God shares His love with us through others in countless ways. He can use creation likewise. A timely breeze during a heart-felt prayer, a parting of clouds to hold back rain, a squirrel darting about the yard to entertain us, and a thunderous stampede of cattle gracefully halting on a dime are just a few of a many ways our God can communicate through creation. *In The Likeness of God*, written by Dr. Paul Brand and Philip Yancey, spiritual metaphors are brought to life to give testimony to our creator in great detail. One of the book's messages is that ours is a thoughtful God who *carefully integrated love into His Creation* for us to experience.

> **God shares His love with us through others in countless ways.**

If turbulence abounds in your life, then you might find a soothing message in a simple cat's purr. If inconsistency or instability plays a part in your life, then the faithfulness of a trusting canine may serve you well.

My son absolutely adores hedgehogs. His term of endearment for this interesting animal is "The Spiky Ball of Death." A friend of the family took one into her care. She didn't expect to be so enamored with it. She said that once you get past the flesh-piercing spikes you could enjoy it! The noise that comes from it was so bizarre that it reminded her of a children's wind-up toy! The hedgehog also reminded her of the wildness and humor that God designed in this creature! It makes us crack up! Our friend introduces children to the creation process through teaching the subject of art. What a blessing she is.

Would a "message" from God's heart to your heart make a difference? What lonely crevice would it fill or what need would it scratch? Perhaps maybe the message is purely cerebral, solving a life-puzzle of sorts, or there might be a matter that weighs upon your heart and is leaving you in a state of pain or distress. God is standing by, just waiting to hear from you.

When was the last time you prayed for your heart? We pray for an open heart to hear His Word or to serve in any way that He requires of us, but do we pray for the *condition* of our hearts? Has it become embittered, timid, or broken? Whatever the state, God wants us at a place that is *near the place* where He originally created us! God wants us to have a content heart. One that isn't choked by the worries of the world[2] or fretting the next day or moment. Remember, God authored our hearts. He created every detail of all of you and me, including a beautiful mind, a healthy body and spirit of peace. So what was God thinking? He was thinking about communicating to us through our hearts!

> Would a "message" from God's heart to your heart make a difference?

The Ultimate Design Challenge!

Only in God's divine wisdom could and entire physiological human communication system be designed which performs the amazing tasks required of us every minute of every day.

From God's perspective, He had to invent ears. A listening system in the brain that not only triggers appropriate receptors, but receptors that permeate to our very core, our soul and heart.

> When was the last time you prayed for your heart?

God invented the listening system complete with an eardrum, cochlea, and nerve-paths wired to the exact part of the brain with little "radar dishes" attached to our heads to catch the sound waves. He even gave us two!

The whole vocal cord noise-making and speech system has just the right amount of tension, attached to just the right place above the windpipe; even the windpipe is an incredible thing. It needs constant lubrication to maintain consistency. This awesome piece of anatomy can create a variety of differentiating sounds, volumes, pitches and intonations. The vocal cords, along with the language forming tongue, cheeks and lips, are gloriously wired to a brain that is language competent and multilingual capable.

> Only in God's divine wisdom could and entire physiological human communication system be designed which performs the amazing tasks required of us every minute of every day.

Every vocal cord comes complete with a range of tones. The user must be able to adjust the tones, without the use of hands—just by spontaneous thought. The vocal cords can create a tone that soothes a child or shouts encouragement at a baseball game. The cords needed to be built to receive subtle signals from the brain that would allow emotions to be expressed through these.

The voice is amazing! Some of the tones and attitudes of expression include; caring, obnoxious, neutral, firm, coy, sneaky, mean, evil, nice, bored, excited, in love, out of love, happy, sad, angry, fearful, contemplative, pensive,

God's Heart in Communication With Man's Heart

interested, curious, encouraging, nurturing and flirtatious. Another feature of the vocal cord sound system is the ability to sing and make music. Not only through talking are we able to express emotions, but also we are able to express ourselves melodically in song! If you have ever been moved or touched by a song or someone's voice, God is the real sound engineer behind it!

Here is a really fascinating part of the noise making system—high tech sound effects! A childhood wouldn't be the same if we couldn't make grunts, groans and coos. As our sound effects system develops we can make a plethora of sounds including motor-powered transportation (like a jet taking off or an old car engine starting up). From cooing at a newborn to speaking foreign languages, our vocal system is unbelievably versatile!

God had to invent lips that have the ability to press-together. Add some humming and a tugboat becomes a speedboat. Go to a high-pitch hum and

> What do you think of a Designer God who equipped us so communication efficient?

we have just added 50 horses of horsepower to a motor! Ok, now lessen the lip pressing pressure and you've instantly changed your sound effect from a speedboat to a motorcycle! When you take a breath making these sounds you've just shifted gears—all for auditory fun and amusement!

Lips are a very important part of our portable sound effect studio. Lips are necessary to form different sounds. What's life without a whistle? Our lips with facial expressions show a fireworks display of emotions: laughing, crying, and smiling. Together with our mouth and throat—which doubles as an echo chamber—we can bond with our animal friends by growling, barking, purring, mooing, quacking, squeaking, chattering and blowing fish bubbles.

Imagine the mind and spirit capable of making these anatomical communication design requirements prior to designing. The sense of humor that is designed into our lips alone is awesome! What do you think of a Designer God who equipped us so communication efficient? Why was He so incredibly detailed in design functionality here? Does He want us to discover some heart truths about Him, as well as ourselves? Perhaps not only is communication important to God but the subtleties of it as well! We can all quickly relate to all these forms of communication!

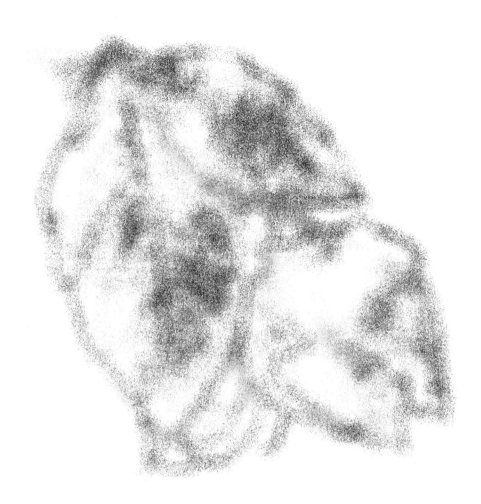

3
Inspirations of God

Where did He begin? There were no palettes, no drawing boards, no paper or pencils. There were no reference books, no "how to" books, no inspiration of any sort to draw upon. To help us understand God's creative process, we need to truly understand that there was . . . *nothing*. There are a number of "design basics" that are commonly used in the world of art to begin a study of God's inspiration and design process.

One of these design points of reference can be seen by examining the spiral found in seashells. This design basic is found in most areas of industry. You'll find it in clothing styles and furniture design. It can be found in the inner workings of machinery, the spring of a watch, and even in our exercise machines. You will find the spiral starting point in an artist's painting. It can be used as the design itself or used to create eye-travel. This contrast-creating element will dance your eye around in a pleasing flow, sometimes simplified in a spiral shape. The seashell has many other facets of design elements including the shells' colors, texture, function and even the sound of its "echo of the sea." They have all been starting points for man's creations. We use whatever works for us, whether it's an acorn, an apple or an ant.

> To help us understand God's creation process, we need to truly understand ... *nothing*.

As we think design through; God didn't even use a collection of molecules, atoms, electrons, protons or DNA as a starting point. Each of these microstructures had to be designed with their abilities to react with one another prior to the elements, creation and function. He designed all these with an empty palette. No chemicals, no organic material to breakdown, no beakers, burners, scopes, measurement devices, computers or even modeling software. If we were to break down nothing to a mathematical equation it might look like:

> God didn't even use a collection of molecules, atoms, electrons, protons or DNA as a starting point.

$$(-\infty) + (+\infty) = (\quad)$$

I can't even use "0" because zero is a number! I am not a mathematician in any way, but even I can understand this.

So from our physical (or earthly) point of view, God's creation starting point is ( ). It was dark and likely very cold (or was it?). The sun wasn't made yet, however we know that scripture refers to God as being "a pillar of light" or "a lamp" or as the "light of the world." In fact, Moses had to bury his head in the crevice of a rock, so as not to be blinded by the light of God's face as He passed by Moses.

These image-packed words from the Old Testament may be deeper than what we read at first glance:

> *How precious to me are your thoughts, O God! How vast is the sum of them! Were I to count them, they would outnumber the grains of sand—when I awake, I am still with you.*
>
> Psalm 139: 17–18

Think of this for a moment. What does it take to generate a single thought? It takes energy, desire, maybe some passion, selflessness and intelligence. Not to mention the physics of it all. Our engineering includes the capability to make the action of thought happen in a single millisecond and then launch it to the precise longitude and latitude of our muscle location traveling at 400,000 miles per second."[1]

This mind-to-muscle "signal" travel efficiency requirement is staggering—and we need this hyper speed communication to catch our balance, steer our cars and to feed ourselves dinner. We can somewhat relate to this kind of speed just as electricity travels through our copper wires. However, the origination source of our household electricity is extremely powerful and strategically harnessed. I know our brains don't store the kind of electricity that our high-powered transformers do, but I certainly wonder what kind of power has the ability to push detailed movement instructions at such high speeds. Our God has a complete and thorough understanding of the molecular!

> **What does it take to generate a single thought?**

In reference to these interesting "thoughts" that David wrote, I asked some math guru friends to come up with the number of grains of sand available in the oceans and beaches. According to Dr. Jason Lyle, astronomer and physicist at the Institute of Creation Research, there are "10^{22} (ten billion trillion total grains, give or take a factor of ten or so."[2] With that number divided into 7 billion (the number of people on the planet), that gives us plus or minus 10%, 1.4 trillion grains of sand per person. What an inconceivable number of thoughts directed toward each of us from God! Yes, this may simply be a figurative use of language, but the ideas are vast and very clear ... Two of which include the fact that God never, EVER stops thinking about us, and second, is that original creation design took into account everything that is required to sustain life. Every single thing had to be thought of and designed! In order for the world to work symbiotically, in perfect balance and order, the number of these thoughts must (and continually do) outnumber the grains of sand!

> Our God has a complete and thorough understanding of the molecular!

> In order for the world to work symbiotically, in perfect balance and order, the number of these thoughts must (and continually do) outnumber the grains of sand!

In order for humans to bring about a single verbal or nonverbal word in any of its forms (thinking, musing, and planning), thought is required. God's Word tells us that God spoke all into existence. No more, no less. Divine pre-creation thought existed before time, it exists today and it is everlasting:

*Before the mountains were born
or you brought forth the earth
and the world, from everlasting to everlasting you are God.*

Psalm 90: 2

God made us in His image and granted us the ability to think and to understand what thinking is. God also gave us the ability to create "lesser things." Because of this, we can get a faith-building glimpse of the "Creator of greater things." These gifts to us are some of what separates us from other life forms and always will.

> *Many, O LORD my God, are the wonders which You have done, And Your thoughts toward us; There is none to compare with You. If I would declare and speak of them, They would be too numerous to count.*
>
> Psalm 40: 5

In the beginning, all things became new. Scripture teaches that God holds all things together. Science teaches that there is a force that surrounds protons and atoms, which is in most that we see. It is invisible, but certainly there. The name that has been given to this is the "Strong Force."[3] Could this have been the inspiration for the "force" that Yoda refers to in the popular Star Wars movies? A more significant question is: Could this "Strong Force" be a spiritual occurrence? Is His power or even His physical presence closer to us than we realize? As Jesus gave the great commission to the disciples in Galilee, He said, "And surely I am with you always, to the very end of the age." Science recognizes the invisible, but can't explain it. This is very similar to the concept of human consciousness; it's there, but it cannot be explained.

> **God also gave us the ability to create "lesser things." Because of this, we can get a faith-building glimpse of the "Creator of greater things."**

Many answers to the questions which perplex us, lie in the invisibles of life. God's thoughts of us lie in the spiritual realm. We can draw on these, but we have to work hard to see them. The visible will be easy to turn to, but we must be willing to take a heavenly approach to see things from God's perspective. Taking this path may be challenging, but it should serve to reaffirm our hope and to strengthen our faith. Many of us have explored much of the visible and have been blessed, but can still feel somewhat empty about many issues. I want to encourage you to explore and give much heed to what is unseen!

> Is His power or even His physical presence closer to us than we realize?

A simplistic example of the invisibles is a mathematical sequence called the "Golden Ratio."[4] It's invisible in theory, bar a few math equations that represent it, but it's very visible in nature. It is an innate, invisible design sense that we all possess (and you thought you didn't have any talent!). Somehow, we can see a natural flow of patterns and can recognize when something is not in the flow of nature. When design is outside of the flow, it can look odd. For example, we've all seen those curly, wrought iron hand railings found in both older and newer home constructions. Have you ever seen the less expensive ones that for some reason, just doesn't look right? They just have a goofy curl or a symmetrical element in the wrong place or the wrong size. Other examples are a hurried floral arrangement or disconcerting dress pattern; they just stick out like a "sore thumb." What makes them hard to look at is that the pattern is out of the natural order of life's design in flow, color, hue, shape, size, texture and so forth. Nearly all of us can pick it up!

> Many answers to the questions which perplex us lie in the invisibles of life.

In a similar way, as we look at creation, we have a desire to know what kind of power and accompanying thoughts are behind all that is seen and unseen.

God's Heart

It's hard to imagine God having any needs, even though the Bible indicates we can grieve the heart of God.[5] It's hard to imagine that He needs our worship. We worship in order to give ourselves an appropriate perspective and heart adjustment. He doesn't need to be encouraged like we do. So what is it that motivated God to call us into existence? Here are a few verses He left with us that offer some insight:

Psalms 37:4 reads, "Delight yourself in Him . . ." As parents, we can delight ourselves in our children. God has given us these parent-child relationships to learn and compare to our relationship with Him. We see our children grow, learn, laugh and cry and our interest never wanes. Through these relationships, it may be fair to say that God's relationship to us in many ways is very much the same. We are His children . . . growing, and learning to look to Him. I can't believe how many times I have repeated to my children, "Would you just trust me?" I am prodded today as an adult with the same admonition from the Holy Spirit . . . In our moments of doubt and despair, He asks us, "Would you just trust me?"

> So what is it that motivated God to call us into existence?

When our hearts are touched, life seems to be complete in some sort of way. Maybe for a moment, we are somehow encouraged. But as time moves on, unfortunately we are prone to drift. Scripture speaks of drifting continually and encourages us to renew our faith daily. With the rush of life we get out of healthy patterns and can easily backslide. Our awareness of God helps us to realize that we have drifted too far from His comfort and we begin to yield to the Creator once again. He reaches back out to us, and the cycle repeats itself. There are so many testimonies worldwide which affirm this process. God wants us to return, He recognizes our frail state. God welcomes our communication, our fellowship with Him. Just as the Prodigal son returned to his father after he squandered his wealth, the father welcomed his son with such unheard of passion that it filled his brother with intense jealousy.

Since we are sons and daughters of a fallen Adam and Eve, life's experiences can keep us from the very essence of why we were created, leaving us dripping with bitterness, anger or resentment. The Bible teaches that bitterness grows like a root. And having an unforgiving heart is one of the major reasons why people frequent therapists. If we wrestle with issues alone, or if we worry excessively, it will choke the life right out of us.[6]

> I can't believe how many times I have repeated to my children, "Would you just trust me?"

> When our hearts are touched, life seems to be complete in some sort of way.

If we are delighting in Him, we have a glad heart. He is lifted up in praise and we are encouraged. As these moments increase, we enter the kingdom in a spiritual sense here on earth. The New Testament teaches us to seek first His kingdom and His righteousness and all these things will be added unto you.

Notice what happens to our spirit when all is right in our relationship with God. All concerns are left behind, even if for just a moment. We are able to laugh, cry or just feel a great sense of peace. When this mini heaven-state comes, it is just that—a very small taste of heaven on earth. The real heaven experience will be so much bigger and better when we pass from this life to the next.

II Corinthians 5: 8 states that to be absent from the body is to be present with the Lord. When Christ was hung on the cross as a sacrifice for our sins, He

> Notice what happens to our spirit when all is right in our relationship with God.

looked over to the thief who was on a cross next to him and said, "Today you will be with me in Paradise." Other verses tell us about a great banquet that will be served—it will be a time of celebration! Once we are in God's presence, life will finally be complete. Our hearts and spirits will know it and we will be filled with unmatched joy.

> Once we are in God's presence, life will finally be complete. Our hearts and spirits will know it and we will be filled with unmatched joy.

No more tiny glimpses. The connection will not fade as it does now, causing us to return to the scriptures, fellowship and prayer to renew our spirit. We will be united with the Spirit of God! If we are able to see past the tests and trials, and by faith accept the free gift of salvation, then we are the truest of friends of God!

> ... *The Scripture was fulfilled which says, "Abraham believed God, and it was reckoned to Him as righteousness," and he was called the friend of God.*
>
> James 2: 23

What inspires God to create? The mere *thought* of us! God desired beings that were made in His image to reflect the kingdom in a fantastic variety of ways! It becomes a sort of "heart art form," to express freely what the kingdom means to you. And it should be an uninhibited expression of what you believe to be the kingdom of God. Are we capable of this kind of expression? God has made us to reflect and represent all that He is. What a challenge God had in mind

> Heaven is in a completely different dimension and we are called to reside there!

for His children; to reflect upon and give hope of an unseen place . . . Heaven is in a completely different dimension and we are called to reside there! We are also told the enemy does not have any destructive or manipulative influence there. The enemy's work and assignments are cancelled! This is truly good news!

In fact, God has many exciting messages which are as unique and special as each one of us. This is one of the reasons He created us; to have millions of individual "heart-to-heart" conversations with each of us. God wants to communicate an extraordinary plan that was specifically designed for every single one of His extraordinary people.

> **God wants to communicate an extraordinary plan that was specifically designed for every single one of His extraordinary people.**

As "creators of lesser things," we also have a message we desire to share with the world. We take great pride upon the completion of a commission or project. We excitedly desire to immediately show our creation to others. We want to touch hearts. We want to talk about it with others. We may have surprised ourselves or perhaps have been truly inspired. Whatever the case, we are proud and want to get feedback. We can't help but wonder if God gets excited about what He created? Is He overwhelmed with pride in His work?

God desires that we come to Him like a child. A child is born with an innate desire to be loved. Once they feel secure and cared for, they are content and happy; we are able to enjoy watching them experience life to the fullest, complete with the freedom to be themselves, unencumbered by constraints of any sort. There are no fears that family or friends will criticize or reject them, their joy just simply happens! Societal boundaries have not hemmed them in yet. A spiritually healthy child may not even know the meaning

> **A child is born with an innate desire to be loved.**

of the word fear. They are thrilled to experience life and live simply—in the moment!

There is immense joy when it comes to experiencing God's presence. In the book of Proverbs, we find that Solomon was filled with a spirit of awesome delight when he walked with the Lord:

> *Then I was beside Him, as a master workman; And I was daily His delight, Rejoicing always before Him, Rejoicing in the world, His earth, And having my delight in the sons of men.*
>
> Proverbs 8: 30, 31

When we approach our Father in complete openness, He too experiences unfathomable amounts of gladness! Perhaps some aspects of the elements He created, such as when a sun's beam pierces the dawn to simultaneously light a cloud's edge and illuminate a mountain's ridge, the rainbow that appears in the mist rising above a thundering waterfall or celebration of light in the constellations, or in a sprinkle of flashing fireflies—all of these are expressions of both wonder and delight. God put all in place to communicate His pride and joy in His creation!

> God has an intensity of passion not known by man.

God has an intensity of passion not known by man and His forms of expressions appear to go on forever, just as His love for us does.

A concept like this can be difficult to comprehend, but God wants us to reveal His love, in its totality. His love has a depth and a breadth and a scope that is beyond our imaginings.

Passion in Action

Here is the challenge. It is extremely difficult in our needy, distracted, fleshy state to look at the time before creation. Let's try to put ourselves with Job as God talks about a time prior to Genesis.

In Job, God says, "Brace yourself like a man. I will question you and you shall answer me." Picture this scene. The Lord is speaking to Job out of a storm. This is not a typical rain shower here. The Bible doesn't use fancy print or interactive graphics to make a point. We have to take it for what it says . . . out of a storm. There are no Hollywood special effects with this storm, lest we forget, ordered by God Himself! This is the real moment in time, a "life drama" unfolding before our eyes. With thunder, *real* thunder, in his voice, God asks Job, "Where were you when I laid the earth's foundations? Tell me if you understand!" In our minds, we hear the horrible peels of thunder and see the white-hot veins of lightening flash across the sky, like a divine exclamation point as God speaks. He continues, "Who marked off its dimensions? Surely you know!" FLASH! CRACK! "Who stretched a measuring line across it? On what were its footings set, or who laid its cornerstone?"

> The Bible doesn't use fancy print or interactive graphics to make a point.

God had warned job to brace himself for this questioning, and rightfully so! The impact of the moment was multi-sensory experience, perhaps overloading all of Job's senses! God's words and nature were firing simultaneously! God doesn't speak out of storms often. God is bringing attention to this point in scripture for a reason.

God is challenging Job and making statements about His glorious creations. Part of the list includes: the morning stars, angels, the springs of the sea, clouds, gates-of-death, gates-of-the-shadow-of-death, the abode-of-light, darkness, snow, hail, east winds, rain, stormfronts, ice, frost, star constellations; Pleiades and Orion, bears and bear cubs, laws of the heavens and lightning bolt destinations . . . And God is just getting started! It continues with eagles, horses,

> Job had lost perspective in his trial. God was trumping Job to make him think; to restore his perspective.

and even behemoths. Chapters 38 through 41 are worth a good reading, just to let all of this imagery soak in, but what is the purpose of this list of powerful, and at times, terrifying images of nature?

Job had lost perspective in his trial. God was trumping Job to make him think; to restore his perspective as a God-fearing, God-trusting, and God-loving man!

We don't have God speaking to us directly out of storms today, but He speaks to us through many other life-storms and trials. The 5th chapter of Matthew states that it "rains on the just and unjust." In the book of James, we find the words, "... when you encounter various trials ..." (notice that it says *when*, not *if*). When we study the literary devices of the writings in the Bible, we can decipher basic biblical concepts which are integral to our understanding of who God is. In Job's trial, for example, God points to the original design and functions of His creations as a testimony to His profound sovereignty and supremacy. He is not to be questioned.

> When we study the literary devices of the writings in the Bible, we can decipher basic biblical concepts which are integral to our understanding of who God is.

God is pointing with a purpose! "Have you comprehended the vast expanses of the earth?" There's a reason He asked Job this and tells us to do this. The next chapter gives us examples of how to discover the heart of God, and see the Creator's design of His masterpieces that keep life fresh, unsolved and mysterious. Comprehending the vast expanses of the earth will help us engage with creation, open our hearts, restore us and give us a new perspective. With this idea as our inspiration, we can become free to create (reflect His glory) like we've never created before!

Really Get It!

Understanding God's passion means that we truly *experience* it. Simply put, He is the Creator with a loving message to us that points us to

Him and we must realize all this was for us to enjoy, like a child enjoys an engaging life event for the first time! If our worship goes beyond routine religious practices to a level of awesome wonder, we are on our way to getting it! If we can look beyond the crisis of the moment, the tyranny of the urgent, the distractions of the enemy, turn the TV off or unplug from surfing the net or texting our friends for awhile, God has an awesome message for us! As we recognize that there may be something bigger than you and me in play right now, we realize that this is a great start and a great place to be!

> Understanding God's passion means that we truly *experience* it.

Job got it. "My ears have heard of you but now my eyes have *seen* you. Therefore I despise myself and repent in dust and ashes." Job states he now "sees," but he is finally seeing with the *eyes of his heart*! Surely an awakening and a renewal of his perspective has taken place.

Perspective

When we lose perspective of God, we are in pretty serious trouble. This loss of "sight" can put life into a death spiral. We are no longer on an elevated plane of existence as He intended it to be. In other words, we are not experiencing the significant connection with God and therefore, we are no longer the beings that He created. We are without our "spiritual compass."

We may not know that our perspective is lost or skewed, and we may just continue on our path, unknowingly headed for dark days. There is sadness and lack of joy and happiness that comes with losing sight of God and who we are in Him. We may chalk it up to the

> When we lose perspective of God, we are in pretty serious trouble.

> The enemy loves to help us along down any destructive path.

school of hard knocks and think, "Life is miserable and I need to learn to live with it." As we spiral down further, our life situations can get worse, compounding upon one another, like freight cars in a train wreck. Unfortunately, our relationships with others can be affected and we can isolate ourselves and push others away. The enemy loves to help us along down any destructive path. Rather than handing us tissues and consoling us with uplifting words, he rocks back and forth on his heels with an evil ear-to-ear grin, pleased to watch as we experience heartbreak, financial, mental and spiritual bankruptcy.

In spite of having a wonderful family or a successful business, or a secure job packed with promotions, or owning our own homes, we can sometimes focus on the negative; that which we don't have. The tremendous amount of discouragement can appear to be overwhelming and the leap from a spiritual death to a physical one doesn't seem that difficult to make—and the enemy is overjoyed.

Some believe that if they bury themselves in doing good; singing in the choir or working on the hostess team during fellowship time, serving as an usher or taking meals to shut ins, they can become whole again. While these are all valuable and worthy pursuits, just serving God alone will not get to the heart of the issue. God's Word tells us that even the religious leaders can get caught in the "good works trap." Jesus stated, "Many of you will cast demons out and heal in my name, but in the end you will come to me and I will say go way I never knew you." Jesus is defining (for the entire world), those that know Him and those that don't. This clear distinction, which exhorts believers to know God (not just *know of* Him, but actually have a *relationship* with Him), prompts us to start checking the condition of our hearts.

With all this, we run the risk of being blind-sided (the enemy is very good at this). We may tell ourselves "All my needs are met and all that I am doing is successful, so what do I need God for?" This seems to be a current trend in our materialistic world, but God desires for us to keep

Him in the very *center* of our perspective, at all times.

We must seek only that which is God's wise plan for our lives. If the kingdom of God is not sought, it will *not* be found. Even King Solomon—the *wisest* man on earth—lost his heavenly perspective even though he had everything His heart desired . . . everything that worldly success and fame could bring! If it can happen to him, it can easily happen to us. Proverbs 30: 7–9 reads, "Two things I ask of you, O LORD; do not refuse me before I die: Keep falsehood and lies far from me, give me neither poverty nor riches, but give me only my daily bread. Otherwise, I may have too much and disown you and say, 'Who is the LORD?' Or I may become poor and steal and so dishonor the name of my God."

Keeping a simple, healthy, God-inclusive perspective is of great significance! Engaging with God's Word and remembering the deeds He has lovingly and passionately done for us will help us keep a right-minded perspective and a God-centered heart:

> This clear distinction, which exhorts believers to know God (not just know of Him, but actually have a *relationship* with Him), prompts us to start checking the condition of our hearts.

> "Will the Lord reject forever? Will he never show his favor again? Has his unfailing love vanished forever? Has his promise failed for all time? Has God forgotten to be merciful? Has he in anger withheld his compassion?" Then I thought, "To this I will appeal: the years of the right hand of the Most High." I will remember the deeds of the LORD; yes, I will remember your miracles of long ago. I will meditate on all your works and consider all your mighty deeds. Your ways, O God, are holy. What god is so great as our God? You are the God who performs miracles; you display your power among the peoples. With your mighty arm you redeemed your people, the descendants of Jacob and Joseph."
>
> Psalm 77: 11

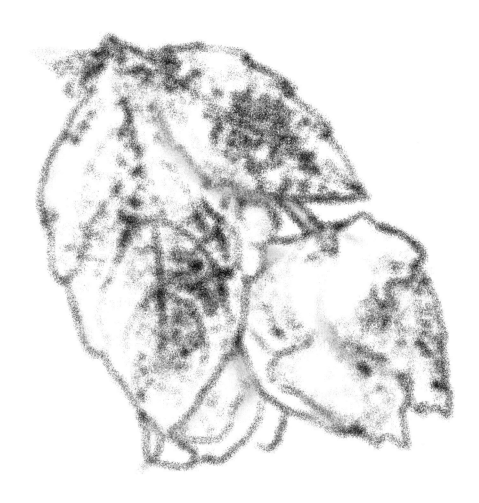

4
God's Creation: Get the Message

When we visit new places, like the local zoo, aquarium, or local park, we can miss the handiwork of God if we are not watching for it. The intent of this chapter is to provide the tools to catch God's vision in creation. If you can catch God's vision, you have experienced an epiphany—a peek into Heaven!

James 3: 17 states, "But the wisdom that is from above is first pure, then peaceful, gentle, reasonable, full of mercy and good fruits, without partiality, and without hypocrisy." The next time you may experience wisdom from above, recognize the influence that it has on your body. Notice what happens to your physiology. When you hear truth that touches your very core, your body wants to take a breath. Not just the usual shallow breathing we are accustomed too, but a full, involuntary deep breath. It's almost impossible to keep ourselves from doing it! This is a shot of wisdom from above and it comes with a side serving of peace. Now on the flip side, information that conflicts with a truth that resides in our hearts, or if someone touches on a personal area that needs addressing, we may have a strong desire to swallow hard afterwards or our palms may sweat.

> If you can catch God's vision, you have experienced an epiphany—a peek into Heaven!

> When you hear truth that touches your very core, your body wants to take a breath.

The next time you feel that you are reacting physically to news or a particular event, ask yourself, "What is God trying to tell me through simple bodily functions?" Whether it's a facial blush, or that sick feeling in the pit of your stomach—take a moment to engage your mind to process what is happening on a physical level and try to think about things on a cause and effect basis. These effects are not a knee-jerk reaction as found in animal instinct, these emotions are connected to our hearts and our passions. They're found in the intimate details of our original design and discovered by us as we "engage" with God. When was the last time that you saw a monkey blush?

Those of us who have children have a great opportunity to show how loving, friendly and creative God is. Teaching children about the design process can give them a sense of wonder and bring the concept of God loving us from a distance to a God who is very close and intimate.

Adults can personally enjoy God's creations depending on each of our unique experiences, educational backgrounds and the insight of our own spirit. We're drawn to our occupations for different reasons. Let's look at creation from an engineer's perspective. An engineer who is blessed with the gift of understanding structure and methodology may gravitate toward exploring structures in creation. Most engineers, with traits of being master problem solvers and a drive to grasp the mysteries of how things work, have a desire to solve structural problems that are put before them. Because of their gifts and talents, they have a natural curiosity toward the inner workings of things. We're not all wired like an engineer, but an engineer yielding to a God-designed perspective should be able to see the creative mind of God in their own way as they explore creation.

A nurse, with gifts of mercy, may see God's loving heart by exploring creations that bring comfort to those who are hurting (something that seems second nature to most nurses).

For example, a nurse may enjoy growing flowers, and he or she may realize the comfort these can bring to people. The unselfish act of giving flowers is an act of kindness, which is so good for those that

> When was the last time that you saw a monkey blush?

need to know that others care. The scent of a rose, the color of the carnation and beauty of a calla lily can all remind us that life is bigger than what we define merely in an earthly way.

An artist may be attracted to creations that are colorful, playful or which offer intrigue in their unusual shape or design. Translucency or texture that is found on an underside of a leaf may be a point of inspiration. Visuals that express depth through perspective—a tree line, or railroad tracks that disappear in the distance—can appeal to the creative soul. There may be repetition in shape, design, asymmetry or perfect symmetry found in a flowering vine or in an unusual rock formation where colors seem to dance together. A composition may be found in the radiance of "sweet light"[1] or on a water's surface reflection.

56   Before the Beginning

I asked a friend, Andy what aspect of his occupation as a landscape architect wows him (I call this the "awe factor.")? He then handed me one of the magazine articles he wrote for an association he belonged to. The excerpt reads,

> ". . . A little two-acre woods and field by my house has something in bloom from March until November and fruit and seed for much longer. What kink in evolution or natural selection caused some trees and shrubs to develop so they would bloom on woody tissue in February or March, others on woody tissue in April or May or June, others on new growth in May, June, July or August, and still others on woody tissue in July, August, September, October, or November? Why do daises and ragweed grow where asters, raspberries and strawberries grow? Why is there something always in bloom, in fruit, or in seed everywhere you look almost all year long? Luck? No, because another part of creation is there to use it. Nothing is wasted."

Another excerpt from Andy, titled "Symmetry by Chance or Design?" reveals a similar rejection of species evolvement ideologies:

> "While looking in an old photo album a while back, I noticed that across several generations of several unrelated families, we were all pretty much symmetrically constructed—two arms, two eyes, two legs, five fingers and toes on each hand and foot . . . Then I was at the zoo checking out our furry and feathered friends in the lower orders of the animal kingdom and I noticed the same thing about them. Legs, ears, eyes, nostrils, fins, gills, etc., all symmetrically arranged! . . . Last weekend I was watching a handful of beneficial insects take aim at the billions of foliage eaters in my yard. Oddly enough, all those bugs, worms, moths, insects, spiders, aphids, ants, wasps, beetles, etc., had symmetry. Even the leaves that were being eaten were symmetrical, and the flowers and the seed pods and fruits . . ."

The symmetry in having two eyes gives us our ability to have depth perception as well as substantially increases our peripheral vision. Hence, participation in sports, driving, and everything else we do is much easier with two eyes! Someone once said that one eye is the backup to the other. The eye is probably one of the most vulnerable parts of our exposed bodies . . . so God gave us a spare!

Symmetry goes deep! We even have brains that are right and left sided! One of Andy's gifts for engaging with God's creation is the deep appreciation and wonder he has for things made by God. Andy continues to question the evolution of the human species by asking the following:

> What makes our physical eyes possible without a concept of vision? . . . That takes a lot more than brain or eyes. It takes a certain light and atmosphere. And, why would an eye evolve? How could the force behind evolution have any foreknowledge about vision and if the conditions were right to permit it?

Many of us are not aware of the God-given spiritual gifts that we already have. We can easily get distracted from using them. Get in touch with your gifts. You'll find a tool or two for perceiving God's creations by engaging your spiritual gifts. There are questionnaires available designed to reveal these to you.

Considering Creation

God tells us to "consider the lilies" He wants us to meditate upon all of His marvelous deeds! But what did He mean by the word "consider"? Webster may be able to aid us in defining this important word:

> **Consider:** To think about carefully; to think of especially with regard to taking some action, to take into account / to regard or treat in an attentive or kindly way / to gaze on steadily or reflectively / to come to judge or classify / to regard / to suppose / to reflect / to deliberate.

We are encouraged to consider the bible; not merely perform a cursory reading of the stories that were so carefully written on these pages, but to *savor* the words of the Bible and keep them in our hearts. It may take some effort to slow down and think deeply about this life instruction manual which God wrote through the power of the Holy Spirit, but you

What was the specific purpose of creation?

will find that it will be well worth it! The messages can come as a whisper at first. As we fine tune our "listening skills," and spend more time with Him, the message will become clearer. This entails patience, prayers and looking at our world with a new perspective. We may get frustrated if we don't feel like God is responding to us right away. Many of us would prefer it if God would just write a few answers in the sky regarding whether or not we should marry that special someone. We might wish that He would send us an email with detailed instructions regarding whether or not we should take that job offer or which university our daughter should attend. But He requires much more from us, and it all starts with learning all we can about our Designer.

As we continue our quest to study what went into creating everything that exists, one important question to ask ourselves is "What was the specific purpose of creation?"

I use a HEAVEN acronym to help spot God's *Before the Beginning* design:

Humor	(Funny bone = healing bone)
Extremes	(Longest, shortest, fastest, slowest, hottest, coldest, hairiest, baldest, lightest, heaviest, weakest, strongest, thinnest, thickest, etc.)
Awe Factor	(It should be *jaw* factor: How far does your jaw drop?)
Variety	(Who doesn't like variety?)
Everyone	(Did the designer have me and you in mind?)
Nuts and Bolts	(Practicality . . . things that either hold things together or are necessary for living, working, or functioning)

Humor

God has a sense of humor. He invented it! Just look at some of those faces of monkeys and dogs that exist. There are pug dogs that look like they tried walking through a glass door. There are animals of every sort with silly features and abilities. God wants to crack us up! Proverbs states "laughter is healing to the bones." And we want those to be in their very best condition, since our bones are the important inner structure of what holds us up! Without bones, we could do nothing! In addition, bones are the key foundation to our individual identities. The bone provides rigidity so we can be protected from most of life's bumps and bruises. It's a place for the muscles to be secured. Our bones create white blood cells so we can fight infection. Along with muscles, tendons, ligaments, organs, etc., the bones give us our mobility.

> God has a sense of humor. He invented it!

If creation makes us laugh, we may be experiencing creation just the way God intended. There is a reason for every living creature on the earth, no matter how odd they might seem.

Monkey Face Orchid (really—just search it on-line!)

One of the most unusual, but common animals is the skunk. Why would God want to make on oversized striped rodent that is armed with its own built-in portable bio-chemical warfare manufacturing facility? One might wonder if God can smell the pungent odor when a skunk gets cornered and responds with a stench- packed stream! The smell can travel for miles, it lingers for days, and it seems to permeate everything! When dogs get sprayed, we douse them in tomato sauce to get rid of the smell. I personally believe it is just an urban legend, since it never seems to work too well (but a real "fix" is 90% Hydrogen Peroxide, 5% dish soap and 5% baking soda). So what good is this foul-smelling beast?

> Yes, even the skunk has a reason for existence.

Yes, even the skunk has a reason for existence. Maybe it's to remind us that really stinky stuff in life sometimes happens. Another possibility is a simple reminder to keep away from stinky stuff ... perhaps it's an analogy to sin, or maybe it's just to make us laugh when we see each other's faces scrunched up in "smell defense" mode! Whatever the reason, when the skunk lifts its tail, it has our complete and unwavering attention!

The nose is able to distinguish ten thousand different smells.[2] Referred to by many as the schnoz, the sniffer, the whiffer, smeller, honker, schnozola and sneezer, the genius is in the placement of this appendage. It is something to be thankful for, in that it's at the helm of our bodies! God even made the nose able to quickly lose a smell in a matter of seconds for those of us who have to spend long periods near tainted odors. Mercy and kindness was thoroughly demonstrated by this thoughtful design feature *before* the beginning! Creation is filled with humor!

What do we do with the frog fish? This fish comes complete with its own fishing pole and it's baited! This is how the fish hunts. A pole comes out of his head while a smaller fish goes for the

> Creation is filled with humor!

bait. The frog fish then snaps up the fooled fish with his "frog-like" tongue! There are plenty of human fishermen and women who wish they had the luck of a frog fish when they cast their bait.

My family and I took a visit to the Cincinnati Aquarium where we saw an animal called a pyjama shark. He literally looks like he is wearing pajamas and dances about half out of the water! It also swims towards the visitors and shoots their faces with water!

At another Cincinnati park, there was a plaque posted at the location where a skeleton of a bear-size beaver was found. It had buckteeth the size of the blade of a garden hoe! Can you imagine coming face to face with one of those? I would be so confused, I wouldn't know if I should laugh or cry!

> God was playful in His design. Just look at the porcupine, the platypus and the pink flamingo!

We need to stop and take the time to experience creation the way God wants us to. See things with a creative eye! Look for the humor! God put these things here for our pleasure! Just as we take pleasure in seeing our kids enjoying each other, toys, critters and creation, I believe that God experiences a similar joy in seeing us taking delight in His creations.

God was playful in His design. Just look at the porcupine, the platypus and the pink flamingo! This playfulness is what the enemy wants to conceal from us. As we remove the enemy's blinders and ask for the help of the Holy Spirit, we will be able to see the joy, the humor and the playful attributes of our Creator.

God's love transcends the constraints of this earthy place. Paul states in Corinthians that, "No eye has seen, no ear has heard, no mind has conceived what God has prepared for those who love Him." In addition to what God has put on our planet to enjoy, He is preparing something beyond what we can conceive. God allowed some of His works to be experienced here on earth, prior to His works in heaven. John 14: 3 tells that Jesus is preparing a place for us. By seeing God's "transcendent playfulness" in His design in nature, we are catching a glimpse of things to come. And that same good-humored spirit is often reflected in our very own nature. God's lightheartedness shines in order to glorify Himself all the more! Here are a few other creations that exhibit playfulness in its design:

> By seeing God's "transcendent playfulness" in His design in nature, we are catching a glimpse of things to come.

- A flying fish, an oxymoron in itself!
- The octopus and jellyfish flash an array of brilliant colors.
- Our feathered friends can come with outlandish headdresses, slipper-like feathered feet and feather designs so unique that they have been used in paintings, fabric designs and even Tiffany lamps for centuries.
- The penguin, pelican and puffer fish all have very distinctive shapes and exaggerated or out-of-place features.
- The night's soothing sounds are like a divine lullaby. The spring peeper's rhythmic sounds, the ribbits and bellows of the toad, and the steady chirps of crickets seem to sooth the soul for a good night's rest. It's even fun to mimic the deep twang of the bullfrog with rubber band "instruments"!

> The night's soothing sounds are like a divine lullaby.

- Hearing a songbird's cheery melody and a rooster's crow are great ways to start a day!
- Polly the parrot does want a cracker! An animal designed to mimic our speech!
- A stick that walks! Finding an incognito walking stick insect is like finding a splinter in a haystack!
- Spots and stripes are all over tigers, giraffes and zebras. Maybe God is working on a checkered cheetah or a heavenly animal

that supports a primitive plaid! In fact, since we can't conceive God's creative work in our current finite thinking state, what He has in store for us might really be different!

And what about the act of play itself? Carefree fun seems to reside in the realm of young uncluttered souls, more so than in the complicated, distracted lives of adults. The purity of spirit that's exhibited when a child skips from room to room, or as they glide through the air on a swing is often perceived as unmatched by other generations. And as we stand at the window, with our briefcase or laundry basket in one hand and a bag of groceries in the other, we realize that something has changed within us. We can't help to wonder what happened to our carefree self. We all have responsibilities that God has placed on our shoulders. On the weekends or on holidays, we try to make time for relaxation and fun, but how do we maintain a playful *spirit*? It's not so much the activity, but the carefree state of a child's mind which fills our backyards and playgrounds with giggles and belly laughs.

When God tells us to come to Him like a child, He is encouraging us to approach Him with simple intentions and honest faith. Jesus says to draw near to Him and He will draw near to us. Jesus is trying to tap into the pure innocence found in children as He tells us to come to Him as a child. In a soothing, Father-like tone, God is telling us, "I know who you were meant to be. I understand who you really are. I made you. I knew you when you were being formed in your mother's womb. I love you child. I have come to give you a full life, a hope and a future. I want to hold you in my arms. I want you under my wings as a hen secures her chicks. I died for you, so every ounce of guilt and shame, all feelings of rejection and self condemnation can be put away for eternity."

> **Every ounce of guilt and shame, all feelings of rejection and self condemnation can be put away for eternity.**

God desires for our spirits to be just as He created them: free and unencumbered by any worldly entanglements, dancing joyously, fully celebrating the intended purity of mind, spirit and being.

> *For you were called to freedom, brethren; only do*
> *not turn your freedom into an opportunity for the flesh,*
> *but through love serve one another.*
>
> Galatians 5: 13

"A keen sense of humor helps us to overlook the unbecoming, understand the unconventional, tolerate the unpleasant, overcome the unexpected, and outlast the unbearable."

Billy Graham, American Clergyman

Extremes

There is a message given to us through all the extremes found in nature. The butterfly is probably one of the kings of extremes. With its extreme molecular metamorphosis capabilities, it can change from one form to another. A caterpillar can make itself into a chrysalis, not over a period of days, but minutes; complete with reflective gold trimmings!

When butterflies emerge from their chrysalis, they have often been compared to Christ rising from the tomb. Their extreme beauty and

Ten-minute transformation from caterpillar to chrysalis.

unbelievable engineering is something to behold. One butterfly, the North American Monarch, has the unique ability to fly as far as 3,000 miles each year to the same precise location!

I want to bring into focus two extreme creatures in specific. The first is our deep-sea volcano dwelling blind shrimp. The other is the ice worm. The blind shrimp resides at the mouth of these volcanoes in water temperatures measuring 700° F. Hydrogen sulfide and methane come out of these vents.[3] Stop and think about this enigma. These heat resistant shrimp exist in colonies very deep in the ocean under tons of pressure. Most meats like chicken, beef, pork and fish cook to the point of becoming consumable at 160° F. Many metals known to man become molten when exposed to 700°

heat. Welders set their torches to this temperature to cut metal. So how can this tiny, but tough crustacean survive in this extreme heat?

> There is a message given to us through all the extremes found in nature.

Let's go for some reasonable assumptions from the perspective of those of us in a thinking mode. A scientist needs to drive for answers. He or she has the tenacity, and burning focus to find measurable answers. These answers need to be in a form acceptable to the scientific community. What happens when the puzzle piece doesn't fit the puzzle, such as the enigmatic volcano shrimp? This delicacy should be served on a bed of rice with a sweet teriyaki blend of sauces on the side! It defies physics, plain and simple. Nevertheless, the scientists continue in their quest, never ceasing to find the answer.

God has wired us with a void that only He can fill. Should we try to fill it with something else, we are fully deceived. God cannot be put in a box and He likes it that way. How blessed we are to have a God such as ours! His non-conforming, physics-busting trump cards keep life fresh and mysterious! And, if we let it, keeps us looking to the original Designer in awe.

> God has wired us with a void that only He can fill. Should we try to fill it with something else, we are fully deceived.

As if that is not enough, let's see what extreme creature we can find creeping along at the other end of the temperature scale.

Can anything thrive in solid ice? Yes! It's our Freon-filled friend the ice worm.[4] According to scientists, there is a three-inch worm that makes its way through glaciers. We can understand earthworms aerating soil, but do we have a clue as to why we would need our glaciers aerated? Furthermore, what about eating, living and even reproducing . . . all carried out while contained in a curtain of solid ice? How can it be and why? Maybe God made it just because He can!—Or maybe confound us and point us to Him.

As we think about more extremes in creation, we come across the awesomely big dinosaur. At the *Answers In Genesis Biblical Creation Museum* in Kentucky, there is a dinosaur exhibit that is fully loaded with realistic animatronics. Hollywood at its best couldn't beat the quality of these high-tech beasts. They roar, snort and one of them even reaches down to take a bite out the next curious visitor! Perhaps it's good that these extremely monolithic creatures are extinct. Life has enough trouble, let alone having to deal with one of these giants roaming our park trails on a bright spring day! The museum also has a planetarium with a rare presentation of the *extreme* size and *precise* order of the universe. A visit to the Creation Museum is well worth the trip!

> God cannot be put in a box and He likes it that way.

Awe Factor

Could simply experiencing *awe and wonder* be what it's all about? God created and we experience it. It touches our hearts and we see *God's order of life!* We smile warmly inside and get a sense of our Creator's handiwork. When this happens, our heart opens up a bit more. As a result, we may attend church or synagogue for the first time. We may have shed our first tear in ten years. Or, we may just end up in a contemplative state of awesome wonder.

The Awe Factor can happen upon observation of a part of God's creation. It involves our senses. We may hear, touch or feel something that was inspired by God. It strikes a chord within our souls. For just a moment, we sense the enormous force it took to create it and our view of life gets strangely bigger. We pause for a moment and we instinctively feel and express . . . *Awe*.

These moments happen to us in individual ways because we were created uniquely. What may cause the Awe Factor in one person may not be the same in the other. The point is for us to be open to them and watch for them. It is through experiences such as these that we can grow closer to whom we were meant to be when God first began to form us.

> These moments happen to us in individual ways because we were created uniquely.

One of the most common ways to experience the Awe Factor is through music. Here in Ohio, a local harp instructor puts on a harp ensemble every year just before Christmas. There are usually 12–15 harpists present. All ages are represented.

The harp itself seems to symbolize peace and tranquility. Artists of old have depicted cherubs fluttering about with harp in hand. Is the simplicity of life's beauty found in a plucked string? David played the harp for King Saul and the king enjoyed the heavenly strings continually, for it comforted his tortured soul.

> Perhaps one of the most common ways to experience the Awe Factor is through music.

When the harp ensemble began playing, the sound resonated to my core. At the far end of the row, a little Asian girl, no more than eight years old, was playing a harp that was nearly as big as her own body. At the opposite end of the row, an elderly woman played her instrument with just as much passion as the youngest player. They were both putting their hearts into the music and it was easy to see that this was their passion.

The magical notes of harmonies rushed toward me and . . . Wham! My eyes began to puddle with emotion. I was caught up in the moment. The incredibly beautiful music carried with it a sense of the love I felt from God who designed us to be able to experience music in this way. This musical journey touched my soul in immeasurable ways. Joy and contentment filled my heart. I breathed deeply and my soul settled

God's Creation: Get the Message 69

into a state of complete rest. At that moment, I experienced a sense of complete awe.

What went on here? Was it a rush of an understanding of what it took to pull this off? Many things were required to make the evening a success, including several different styles of harps, such as the Classical and Celtic. Was it an acknowledgement of the man hours and care it took to make each harp? Maybe it was the timbre each harp produced or the sound of all of them playing simultaneously.

What about the years of experience the instructor has, the passion and focus she had to keep doing what she believed in? I considered the hours of practice and lessons each student took to get to the place where they could now play the songs that had been so meticulously selected. There is no doubt in my mind that something beyond this world took place as I experienced this moment, and it was indeed wonderfully overwhelming! I can honestly say, I don't know what exactly happened. Analyzing the experience wasn't important or necessary. I was just simply touched and filled with awe.

> There is no doubt in my mind that something beyond this world took place as I experienced this moment . . .

The Awe-Inspiring Universe

Our universe is skillfully described in a book entitled *Taking Back Astronomy, The Heavens Declare Creation,* by Dr. Jason Lisle, astrophysicist. Dr. Lisle makes several observations that prompt grand wonder. For example, to help understand the distance between the earth and the sun, Lisle estimates that it would take a person traveling at 65 mph, 163 years to reach the sun. That distance is just the right amount from the sun so we don't burn or freeze. Are all of these amazing celestial arrangements merely accidents? In addition, a drive to Pluto would take 6,500 years at 65mph![5] Were these starry night-lights put into place just to steer our ships or is there another deeper meaning waiting to be discovered?

There is so much that Dr. Lisle brilliantly illuminates. The sizes of the sun and the moon are spatially the same (angular size).[6] The moon and the sun's diameters just happen to be the diameter of a common writing instrument. If both were depicted in a painting or photograph, they would measure about a ¼" in diameter. Can you believe that something so spatially small is able to give us enough light to see, grow our crops, warm the waters and fully sustain life as we know it?

Can you imagine the design challenges that faced God as He sat behind His drawing board of life? Hypothetically, was God scratching His giant head and thinking out loud? What would He be saying to Himself? Maybe it went something like this:

> At that moment, I experienced a sense of complete awe.

> "OK, I need a life-and-light-giving source—let's call it the sun—and we'll use hydrogen as its fuel. It should perpetually burn to create heat and light that should radiate the right living conditions about 94 million miles from the earth . . . No, on second thought, that is a little too far away and my creations might get a little too chilly. 92 million miles away would be devastatingly too hot, so let's put it at 93 million miles away; that it should be just about perfect. If they choose, my creations can move a little bit south to feel more warmth from the sun. I hope they can feel my love for them in my careful creation of the sun and the placement of all the planets, moons and all of the stars in the sky. I have created all of this . . . for them."

God's careful loving thoughts create again in me, the Awe Factor!

Variety

Most of us like variety. Because of this, we can purchase miniature shrink-wrapped cereal boxes, 31 flavors of ice cream at Baskin Robbins®[7] and hundreds of channels in our cable packages! We enjoy diversity! To quote a popular motto: "Variety is the spice of life!"

God created variety in everything! As Paul Garner shares in his book, *The New Creationism*, we have about "1.7 million species of plants, animals, fungi, microbes and other forms of life that have been identified and named by biologists, but estimates of the total number of species on Earth vary greatly from ten million to 100 million."[8]

> Are all of these amazing celestial arrangements merely accidents?

Pick your favorite creation and think about it for a minute. Let's take flowers for example. The variety and transitional colors of floral species are extraordinary. Color in and of itself gives us energy and excitement! Color on a flower decorates our fields, our homes and our walkways. Attach a flower to almost anything, and it will cheer the space right up! Give flowers to someone who has done something nice for you, or give a plant to a new neighbor or someone who is recovering in the hospital, and it will brighten their day!

Isn't it marvelous that God made our brains flower-reactive? When you get down to the science of it, when we see something that evokes pleasure, our brains produce endorphins (happy chemicals that permeate to other parts of our mind). The sight and scent of a flower can make us feel very good, if not loved. When they are picked they just need a little water. They're lightweight, so we can take them anywhere. (If a flower

> Can you imagine the design challenges that faced God as He sat behind His drawing board of life?

doesn't work then maybe a twelve point buck sighting on a hunting trip or the smell of a 10 oz. sizzling steak on a grill will do the trick. We are wired to be "reactive" to different things!)

Isn't it marvelous that God made our brains flower-reactive?

The softness of a pedal is unlike any other natural texture. The variety of colors, scents, sizes, textures and functions that flowers come in is staggering. In fact, one variety of lily offers pollen that attracts only one bee on only one day, the melipona bee. Only this particular bee knows the secret to pollinating this plant and without it, we would have no vanilla bean![9] Thank the Creator that this happens, or we would certainly have to survive with fewer varieties of cakes, cookies, pies and other desserts, for we would have no vanilla.

Our brain is designed to highly welcome variety. Our miraculous nose, the wiring to the brain from the nose, and the brain itself can in a brief sniff, interpret the difference between the scent of a rose and the scent of a carnation!

The Great Artist provides us with an abundance of visual multiplicity in His work. At the right is a visual that represent the variety found in a small specie sampling of our feathered friends. The detail, design, construction and functionality observed in a simple quill, can leave one without words.

The butterfly offers lots of colors, shapes, sizes and a wide assortment of species. The Owl Butterfly is so named because the patterns on its wings resemble the eyes of an owl. In fact, the white-feathered color on the tips of the butterfly wings makes the entire body of the butterfly look like an owl's head. We understand that the practicality of this coloring is for camouflage and serves as a defense mechanism to scare away predators. If you look

The Great Artist provides us with an abundance of visual multiplicity in His work.

God's Creation: Get the Message 73

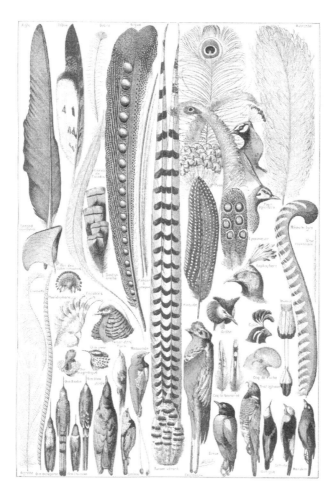

Plumes

at the wings in the spirit of the Awe Factor, it can become an artist's canvas, and we can easily see the Designer's skill; pure brilliance. There are well-conceived patterns, balance and shear symmetry of design. Elements of color, texture and shape create interest and intrigue to the observer.

Look further and you will discover a divine intention of fine workmanship. There is a 3-D quality to the pattern and colors. Warm and cool colors are used to visualize depth and pleasing coloration. The fine markings, as if someone is using a small sable-paintbrush, are found within the elements to illustrate form. If you look closely, you

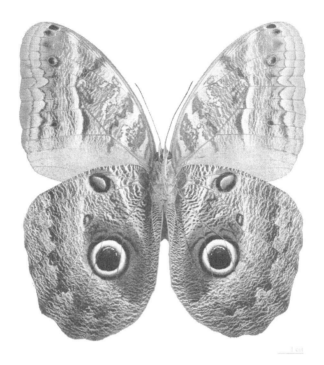

Owl Butterfly

will see fine color strokes, like that of a paintbrush, that create the illusion of highlight and shadow. Each stroke is needed to show form and the glassy look of the "eyes." Even the depths of eye sockets are depicted! The wing's canvas-like surface is not overworked as a beginning artist might do. To look at this work, you would think that the artist was relaxed and enjoying himself, and knew exactly what He wanted to accomplish.

God designed the Owl Butterfly to frighten away its predators. But I believe He also designed this handiwork to be appreciated by His people from continent to continent as it makes its flight! Its intricacy and beauty are not to be viewed on a gallery wall. This amazing painting is on the six-inch

---
God created the colors found on the Blue Morpho butterfly without pigments!

God's Creation: Get the Message

winged-canvas of what some may consider an insignificant creature—a butterfly!

> Are we beginning to experience an Awe Factor yet?

The following picture is the South American Blue Morpho Rhetenor butterfly with wing surface sections under different magnifications. God created the colors found on the Blue Morpho Rhetenor butterfly without pigments! It looks so vivid, as if someone had dipped its wings in cans of bright electric blue paint. God used light refraction and light deletion to create color!

Wow! (Are we beginning to experience an Awe Factor yet?) All that work and preparation for just a few weeks of flight! Realize that just a few weeks ago, our butterfly friends were just a translucent ooze transforming from a caterpillar to their new state. Why all this fuss? Assume for a moment that each one of these is indeed an artist's canvas. These "art pieces," in their splendidly vast variety, do get some great exposure, but the work is unsigned! Maybe we are to acknowledge the artist's identity on faith and offer again our uninhibited applause, which is what faithful followers call praise.

Another thought demonstrated in our butterfly magnification is the order and intention of design. What keeps this tiny creature suspended in flight? And what would we see at 200,000 times magnification? It seems as though just as the perception of eternity can reach past the celestial, a like perception of eternity can be found in the very, very small. It just keeps going on forever! ...

> It seems as though just as the perception of eternity can reach past the celestial, a like perception of eternity can be found in the very, very small. It just keeps going on forever!

It's all designed and managed, whether it's a creature on land, in the sea or in the air.

76 Before the Beginning

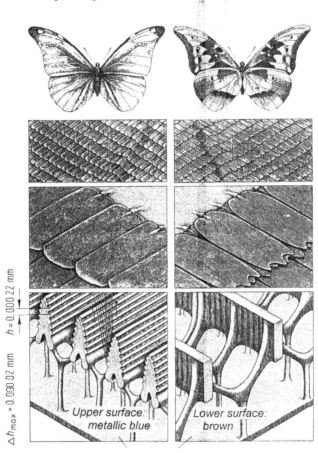

Morpho Rhetenor Butterfly
Magnification: 20,000 times[10]

Form, function, beauty and wonder can be found everywhere, all intended for you and for me! For certain, God did not want us to get aloof about His creation. How could King Solomon get so disinterested with life? As kids we would whine, "Oh, I'm so bored!" My personal opinion is God had to come down here Himself and tell us to "consider," and remind us not to forget His works of old!

Everyone

All of creation is designed to grab our attention and make us think, to cause us to look to Heaven and to help us

> God loves us and is passionately pursuing us.

God's Creation: Get the Message

to get the fact that God EXISTS! By immersing us in the vastness of God's creation, He has a motive in mind: love. God loves us and is passionately pursuing us. All that we see was put here to sustain life; your life and my life, and to demonstrate His abing love.

God gives us time to make choices. Some are good and some are bad. Not many bother to define these choices. In America, we pursue the American dream aggressively to the best of our ability. If we chase this dream in excess, it's easy to miss the messages given to us along the way. What's important to God should be important to us. After all, He did travel to the ends of the galaxy to name the stars and set all in perfect motion!

> God has shown through history that He is a very jealous God.

> The first is to acknowledge God through a simple prayer—a prayer from your heart.

Other hypothetical questions that God maybe asking Himself . . .

"Will the adults finally see themselves for what they are—my children? Will they be able to see themselves as a child that can't find the juice box . . . sitting front and center on the eye-level shelf of a refrigerator? The adults will even enter a state of utter-bewilderment because their children can't see something right in front their faces."

Perhaps God looks down, scatches His chin in wonder and says...

"I put my children right in the middle of my billions of creations, gave them an intellect and five highly sophisticated and extremely sensitive senses to explore these creations and you'd think that they could see their Creator! It cracks me up every time someone finally gets it! Did I give them a little too much of my "know it all" attribute?"

On the more serious side, some will reject or rebel against God for a number of reasons. Some think it's not a convenient time to get involved and pursue the things of God. They may be concerned with the opinions of peers, non-believing friends, family or spouses. Some sincerely believe they are following God, but spend more time in front of the TV or on the Internet than they do learning about God and His Kingdom. God has shown through history that He is a very jealous God.

If you are not following God but would like to, may I suggest two things. The first is to acknowledge God through a simple prayer—a prayer from your heart. That's how we talk to Him! He tells us that He hears our prayers! The second is to contact a friend that is attending a Bible-based church and tell them of your interest. If you are unfamiliar with the churches in your area, use a search engine to search the topic or use the phone book and visit a local Evangelical church. I suggest Evangelical because most churches in this vein have a strong foundation of outreach and give a direct and purposeful message of the Gospel.

Finally, according to the Bible, we're not to know the hour of Christ's return, but we can know the season—and many believe the season is near. Try not to miss any opportunities to pursue God and His Kingdom!

Nuts and Bolts Practicality

God is so practical it is unbelievable. What's unbelievable is how easy it is to take these creations for granted! We moan and groan about trivial issues, and get down because someone at church offended us. We can be absolute brats! Our family could be in great health, the kids doing great in school and we have a roof over our head, and still we are not counting our blessings and rejoicing continually. So what's our problem?

First, understand we are in a fallen world . . . Adam and Eve chose wrongly. The effects of sin are everywhere. However, as we look at how things were back in Garden, before Adam and Eve were tempted, the

Garden was as God intended. God says to come to Him like a child. He promises to give us peace in any circumstance. He also tells us to, "... seek first His Kingdom and all these things will be added unto you." This is pretty awesome. He gives us a promise that His Son Jesus is coming again to reign.

We lose perspective because of the fall of man. He has given us the tools to get it back.

Even in tough times, God is practical.

Look at creation through the "practicality filter!" Pick any area of creation that interests you and observe it through the lens of "creation justification"—*why* God made it *that* way. Let's look again at the stars. The stars help to steer our ships. The moon gives us a nightlight. The planets are benchmarks to measure seasons and time. How practical and orderly of God!

> We are in a fallen world ... Adam and Eve chose wrongly. The effects of sin are everywhere.

EVERYTHING was planned; from the celestial universe to the universe of our bodies. Our skin, for example, is very practical. It has several purposes. It gives our delicate organs a place of safety and holds us together. It makes us easier to look at (which helps in finding a mate), and the skin protects us from danger. The skin is designed with very practical receptors. These help to keep us from getting too close to a fire and tells us to get out of the cold.

Our skin protects itself from repeat use by developing calluses. It gives us our individual identities. It comes in different colors, it changes color locally or overall to tell us there is a problem. It changes in size as we grow larger and it stops growing when we get there!

God gave our skin dynamic growing instructions. These precise instructions tell our skin what parts to close up and what parts to leave open. Our noses, mouths and ears allow entry for smells, food and sound. Our eye sockets provide a place for our eyes to view God's glorious world! Our skin provides the envelope for the *beauty* of our form to take shape. I think it's very nice of God to outfit us in a morphsuit; a literal skin onesie!

The skin must regenerate in order to heal itself, so repairs can be made and injuries healed.

The skin is touch-sensitive. You can sooth a soul with gentle touch. You can offer security, warmth and comfort. The muscles underneath it give it form, grace and beauty.

Hair grows from it for many reasons. We can keep extra warm by letting it grow long on our

> Catch the Godly passion and love exhibited through design. Disregard assumption-based ideas about Godless creation origins that distract us from truths about our Creator.

heads or face. It conveniently regulates our body temperature. And, since there is such a variety of hair colors and styles, it helps us to have unique identities from one another.

Whatever makes your heart glad in Creation, take it through the practicality filter anytime and it is guaranteed to help change your countenance! God was designing and thinking of us from His very first thoughts of creation. Catch the Godly passion and love exhibited through design. Disregard assumption-based ideas about Godless creation origins that distract us from truths about our Creator. These human, finite concepts are merely passion distractions.

It is good to take the time in prayer and thank Him for design and creativity. The passion God has for us shows us that He wants us to thoroughly exalt His creation and experience life in every sense of the word.

> ... They blessed the king and then went home, joyful
> and glad in heart for all the good things the LORD had
> done for his servant David and his people Israel.
>
> I Kings 8: 66

> *Through him all things were made; without him*
> *nothing was made that has been made.*
>
> John 1: 3

 The Creation acronym HEAVEN is simple and easy to memorize. Perhaps you can create one of your own. We can easily spot HEAVEN in what we see. In our comings and goings, we should make a little extra time to explore books on favorite interests, nature, zoos, parks and gardens, and look through a telescope or a microscope. Life can get richer very quickly when we look at things a little closer! Engage with God's many creations in ways you understand and can appreciate. When we get the "message" of creation, it would be good to remember to compliment, thank and praise Him for these things. These are times to express ourselves in pure joy just as a child giggles freely without any inhibitions. God's joy will change your heart and other's hearts as you share your discoveries!

 God gave us a proverbial candy store filled with a multitude of flavors to discover—if we pay attention to our curiosity! Explore and investigate, dissect and then piece back together, take apart and make connections. Next: combine, mix, agitate (meditate), water it, give it the acid test (or whatever test you can dream up), electrify it, shake it, stretch it, taste it, record it, play it, float it and fly it into outer space. You never know where God will help you to land, but one thing is certain: It's going to be awe inspiring!

5
Diversity and Devotion

God was very intentional on the subject of diversity. Diversity covers the categories of creation while variety differentiates species, grouping or family. The big picture is stated in Genesis. The birds of the air, fish of the sea, animals, plants, stars, sky, sun, air, water, man, woman, earth and its natural resources, and energies are fully complete with many mysteries.

God didn't give us a gigantic tennis court floating in the cosmos to reside on, with evil on one side of the net and good on the other side.

If He did give instructions, God could have left a single-line instruction sheet that read, "Stay clear of the edges," but He didn't! He left us a highly creative, humorous, extremely amazing universe filled with mind-blowing variety, practicality and diversity, for everyone to enjoy! For an instruction manual, He left us the Bible—the inspired Word of God. The detail covered in the Bible is astonishing. It's not always cut and dry. We may think it is because of our worldview and limited wisdom. It could be deliberately designed that way to spur on debate, discussion and challenge us in our faith.

> *And without faith it is impossible to please God, because anyone who comes to him must believe that he exists and that he rewards those who earnestly seek him.*
>
> Hebrews 11: 6

Diversity can be found in the make-up of our character. Our moods, temperaments and attitudes can sometimes get us into trouble. However, on the bright side, could these be instilled to push us to the end of ourselves? Perhaps they push us past the familiar into dependency. For example, when we are impatient, the frustration can trigger a disgruntled mood. When we find ourselves in a situation that is beyond our ability to solve the discomfort can push into the face of God. Many of us have been there. When we are pleading for an answer or relief, this comes from the heart. That's right where God wants us.

In Ecclesiastes, King Solomon's countenance is revealed. Solomon, who was the son of King David, was blessed with every material possession possible—everything imaginable—including wisdom. His kingdom stretched from the Euphrates River

> God didn't give us a gigantic tennis court floating in the cosmos to reside on, with evil on one side of the net and good on the other side.

> Our moods, temperaments and attitudes can sometimes get us into trouble. However, on the bright side, could these be instilled to push us to the end of ourselves?

all the way down to Egypt! However, his quest for wisdom eventually drove him to despair. But in the end, God used Solomon to direct His people back to Himself.

King Solomon had everything a human could ask for; from cattle, horses and servants, to vast property and monetary means. But still, he searched for more. He was looking for purpose in life—the secret of existence. Solomon finally asks, "What is the meaning of it all?" Solomon surmises that all is meaningless! In Solomon's eyes, man works and then he dies, the sun rises and sets . . . redundantly. Even the natural elements, like the life-giving circulation of the air (winds) and water (streams) constantly repeat. This prompts him to complain that all things are wearisome! Remember, this King has it all, including wealth beyond anyone's wildest dreams! But it would seem that this wise man has quite a dreary outlook on life! The bottom line is, this king of the immense land of Israel, is quite bored! He shares that he is not fulfilled, adding that there is ". . . nothing new under the sun." Solomon bemoans nearly every other aspect of his life, as if to say why do we "bother living at all?" Solomon has become a supreme pessimist; he doesn't live. He merely exists—and does so without any dreams, without aspirations and is without any anticipation for what the future can bring.

> Solomon has become a supreme pessimist; he doesn't live. He merely exists.

However, as we explore God's creation, we will discover that God's Word will give us hope! First and foremost, there is a loving and passionate God who is building a place for us to go after this life. God's Word helps us to get past the clutches of the enemy and to avoid Hell.

We can immediately have peace of mind and heart! Are not these things of God good? Is there anything on earth more valuable than our souls? The discoveries that we are loved and cherished should give us a glad heart full of joy and a broad grin! Is that not worth living for?

> Somehow, in his reaching for the stars and seeing life in his own way, Solomon lost sight of God. Does that sound familiar at all?

Solomon had a load of earthly possessions, but one thing he was missing was *perspective*. Somehow, in his reaching for the stars and seeing life in his own way, Solomon lost sight of God. Does that sound familiar at all? Solomon chose to run after earthly wealth (and other Gods) and neglected to recognize the diversity in the beautiful world that God had created. Solomon did not have a grateful heart.

We do not know if Solomon fully recovered from his lost ways. However, he did point back to God by instructing his readers to "Fear God and Keep His commandments."[1] Did God use Solomon's temperament, attitude and moods to come for His own glory? Even in our disappointing and selfish human nature, God can bring us closer to His heart.

Solomon's example was quite personal, but hopefully still beneficial. God certainly shouts, "I Am" through the make-up of our character.

Words cannot even describe the scope of diversity that our Creator put on our planet.

We have poisonous and edible-berries, soft and hardwood trees, cloudy and sunny days and we have changing seasons. We have many diverse-energies in reach to draw on; coal, oil, wood, water-electricity, gas, oxygen, wind and the sun. In addition, He has gifted many people with diverse talents

> God is motivated purely out of the incredible love He has for us.

in the arts and different areas of specialized knowledge. Thankfully, He has blessed many with spiritual gifts such as discernment, mercy, hospitality, an affinity for children, leadership or musical abilities.

So how can we shape our thoughts of diversity into a simple understanding, which brings glory to God? If we can recall that, no matter what we are considering in creation, there is an active, loving Designer (God) behind each creation and that God is motivated purely out of the incredible love He has for

> God rejoices that His children can travel through this earthly dimension to a spiritual one and dock our souls to the Alpha and Omega.

us. He wants to get our attention, show His greatness, and warm our hearts just enough to bring a smile.

God rejoices that His children can travel through this earthly dimension to a spiritual one and dock our souls to the Alpha and Omega. If you can acknowledge that one of the smallest specs of creation, such as a pollen particle, is a design of God's, it is good. We have only touched on a small aspect of diversity, but we can surmise that it would take commitment or an incomprehensible *divine* devotion to design the *diversity* that is present in much of our daily lives, as well as into life itself.

Devotion

Webster defines the word *devotion* as religious fervor, an act of prayer or private worship, a religious exercise or practice. A second meaning is an act of devoting—devotion of time and energy, or a state of being ardently dedicated and loyal. The third meaning has to do with the object of one's devotion.

Some of the synonyms include: adoring, affectionate, loving, fond, tender and tenderhearted.

We are just looking at the second and third meanings. The example Webster gives here is, "a devotion of time and energy." While Solomon faltered in the areas of being dedicated and faithful, creation provides a vast array of examples of devotion for us to consider. The Emperor Penguin, living off the coast of Antarctica is the epitome of familial devotion. They were created with a designed instinct to protect their eggs to a point of sacrifice which is difficult to grasp. Once the female lays the egg, the male rests the egg on his feet for over two months while the female goes off to sea to feed. The male waits in subzero temperatures and uses its time and energy to keep the egg warm. It does nothing else! The males huddle together and constantly walk in circles to generate body energy and to protect the new life from bitter temperatures and harsh Arctic winds. The male doesn't even eat during this time. Finally, after nearly 60 days, the female returns and shares her bounty.

In the same way, God asks us to consider another bird, the sparrow. He tells us He feeds them. Why does God care for these birds? After all, it's just a little bird, one of billions on the planet. But God is Holy and perfect in His care giving. God tells us not to worry because if He is caring for these little birds, He certainly is caring for us. God indeed is devoted to us; we are children of the King!

Our Heavenly Father has wired us for devotion! We can exhibit devotion to each other, to our children, our families and to Him! He has a handle on this inner wiring and it's fair to say that His devotion to us is *extreme*. He has done His part in all His life sustaining creation.

What intricate complexities it takes to be omnipresent; taking care of our physical, emotional and spiritual requirements! He wired us with a personal will, which grants us our individuality—our freedom to freely choose our eternal destiny. The engineering and planning were the important, rudimentary activities of laying the foundations of life. As earthly designers with human limitations, all we can understand is that thought is required prior to creating, before the work of "creating" commences. But there was so much more. In fact, we may never see the entire picture until we reach Heaven.

We're made in His image and He has given us the ability to understand thought. Scripture is clear that His "thoughts" of us outnumber the sand... this indeed is the first part of creation. We should be satisfied with that. It gives us something to look forward to when passing from this life to the next: learning how God converts thought into tangibility.

In a past TV series called *Amazing Stories*, author and director Steven Spielberg made a valiant attempt to demonstrate an act of "creating." One particular episode was titled "The Mission" and it included a very exciting rescue scene. During a nasty gunfight, a World War II ball turret gunner and aspiring cartoonist became trapped in the glass bubble under the plane. In the course of the battle, the gun turret entry hatch had become fused shut and the landing gear on the wings were blown off. The men finished their mission and headed back to base to land.

> As earthly designers with human limitations, all we can understand is that thought is required prior to creating, before the work of "creating" commences.

The plane was running out of fuel and the men had to land. Everyone knew what the outcome would be with no landing gear. The gunner had a brand new family waiting for him back home. He had his sketchbook with him in the gun turret. As the plane made its final approach, Spielberg allowed the viewers to see that the gunner was thinking about his family and tears were flowing down his face. It was a very emotional scene. Then, in his desperation to survive, he sketched feverously the WW II plane in which they were trapped. Then he added the wheel and landing gear where it should be on the wings. The camera zoomed in as he took his foundational sketch lines to expressive dark, heavy lines giving the landing gear its form and definition. He drew harder and faster, crying with passion as he worked, knowing somehow his life depended on it! Then magically, the lines on the paper were emerging under the wing of the plane and formed the complete landing gear just as he sketched it. Just then, the plane touched down and the drawing held until the plane stopped and they were able to break the hero free from the prison of his gun turret.

> There is a similar passion and *devotion* in mind for each of us, just as He "imagined."

The gunner's sheer devotion to his family, his determination, desperation, belief and his powerful imagination gave birth to life saving "reality" in this colorful work of fiction. However, with God, the details, the symbiosis and marvel of it all, there was a similar passion and *devotion* in mind for each of us, just as He "imagined." We are *skillfully wrought* in creation.[2]

Each creation is complete as He envisioned.

When it comes to creating something out of nothing, we can't cross the lines of emergence (the process of coming into being or appearing), no matter how hard we try. We are drawn that way, just as God intended. God even engineered people to be "devotion" capable. Let's try to put that in an artist's sketchbook or on an engineer's blueprint!

> God even engineered people to be "devotion" capable. Let's try to put that in an artist's sketchbook or on an engineer's blueprint!

Our wiring and engineering, and everything we need on this life-sustaining planet are complete. That includes our emotions, thinking process, and even our ability to make decisions. The extraordinary devotion God has for us is very apparent in the details of our design and creation's design.

God continued His devotion to us and gave us millions of plants. (A garden, with a lower case "g.") This garden offers us beauty, oxygen, energy and food.

The "garden" is a day to day reminder of the indescribable affection and passion that God has for us.

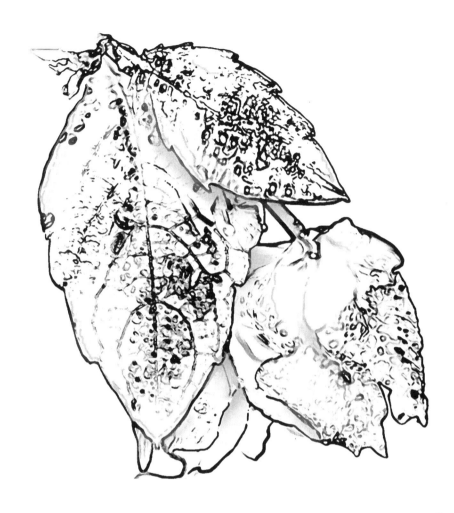

6
The Garden Around Us

There are two gardens, the Garden of Eden and our own "garden" around us. The Garden of Eden is our benchmark of things that were, and of things to come (a new earth in place without the devil's influence). And isn't that an exciting thought? In short, the Garden of Eden was God's intended way of life for us, free from curses and sin—a paradise. Adam and Eve made a choice that represented their independent and sinful nature. Unfortunately, it earned them a one-way ticket through the exit door and into the unknown. Before you bemoan the actions of Adam and Eve, remember that we probably would have made the same decision if we were there! God knew this and the rest is history.

When the books of the Bible were written, they were written by inspired men of God. The people of the time worked the land and sea with their hands and very basic tools and resources. If a local newspaper were printed, a sample headline would read . . . "Since the advent of the plow and sickle . . ."

> The Garden of Eden was God's intended way of life for us, free from curses.

But before their fall, Adam and Eve didn't need tools. And even if they did, they wouldn't break down! God didn't require them to toil by working the land for sustenance. The land was not yet cursed. There weren't any plant destroying insects or greedy gophers, misdirected deer, blight, disease, weeds, Canada thistle, frost, floods or drought to worry about.

There was no warring against the elements and invaders in the Garden of Eden. They picked and they ate! The atmosphere was like that of our humid greenhouses. Imagine how easy they had it! There was perfect order with each other and with God. God was even walking around with them and they knew the sound of His steps.

> *Then the man and his wife heard the sound of the Lord God*
> *as he was walking in the garden in the cool of the day . . .*
>
> Genesis 3: 8

Even the animals had a pleasant disposition. The wild wasn't "wild" yet!

If we were to go back before the Garden of Eden, before the beginning . . . and look at the Designer's drawing board of plant life, it would be difficult for us to even glean the incomprehensible depths of the creation process. As one who visualizes, I was experimenting with a photo I took in a garden. The result of one image was somewhat accidental and astonishing. The leaves of the plant looked as though they were in a mixed designed and engineered state of creation. Shades of green appear to be just coming into existence. The color blue has long been symbolic of water or new birth. Water droplets in line form seem to represent a vascular pipeline for controlled water distribution which is critical for its existence. The intense oranges and reds look like an energy map of sorts needed to transform nothing to something. The leaf veins seem to appear as a foundational skeleton. Even the black represents the Creator's starting point on the leaf. The photo illustration works perfectly as a representation of *Before the Beginning* thinking—a perfect choice for the cover of this book.

With each design, God had a separate mission in mind. One of the missions of the Garden was to provide the body with fuel. The body will need proteins, sugars, minerals, fiber, and nutrients to run at peak performance. Again, remember, there was no stock pile of raw elements to draw from; no science lab supply store to order from. God brought the vision straight from His mind to life and created everything He desired Himself. He generated all that was needed.

In order to go deeper into the subject, let's imagine that God has a sketchbook and some handy test tubes. Before the beginning, God might have said, "The fruit needs to be portable, because my people will be transient." So, He begins to sketch an orange. His next thought may have been, "I'll give a protective outer coating. I will call it a rind." (See the photo/scan on the back cover). "Each orange section should fit nicely within a child's hand. The thick rind and central column will insulate and feed the fragile water cells inside. The nutrient-dence cells will burst out with flavor with each bite. All the fruiting plants and trees will need to be established where people live." Next, God begins to sketch individual seeds in patterns recognized today. "Each ripened fruit should

> **With each design, God had a separate mission in mind.**

grow and contain plenty of seeds. The seed should be easy to harvest, store and spread around. It needs to have three key control elements to begin sprouting: water, soil, and sunlight. Without these things, the seeds will remain dormant and be able to be stored for long periods of time."

God initiates a close-up cut-a-way drawing of a soil layer and in the background, a sun and rain mix, like our combination sun and rain showers that we sometimes see today. Overlaying everything are dashed lines, representing the elements. His pencil begins ghosting multiple colors out softly, like an airbrush. Some patterns are random and some are orderly. Perhaps the Master Gardener has indicated an invisible "radar" which surrounds newly sprouted seedlings, almost like a plant-to-insect signal is being engaged.

> The Master Gardener has indicated an invisible "radar" which surrounds newly sprouted seedlings, almost like a plant-to-insect signal is being engaged.

The drawing breaks out into a vast number of translucent layers as the stack fans and rotates simultaneously. Earlier drawings made, rise up in a methodical order. One layer has a familiar DNA strand on it. Branching off of it is a series of neutrinos. A sub atomic particle which has an ability to move through matter, these neutrinos seem to move in and out of existence. Another layer has honeycomb illustrations on it with a myriad of equations surrounding it. One drawing is of a bee. With wing, leg, and antenna entry points and tendon-like strands holding all together. It has arrows indicating blood flow direction. On another, an animation of the bee's heart chambers opening and closing. Gauge-like settings measuring gravitational forces and more equations are on the next layer, along with measurements of wall thicknesses, wing speed settings and more ordered math symbols.

Another layer has Morse code like lines mimicking the task of a bee, a blossom laden with pollen, the airflow, and the pressure created above and below each wing.

God continues His work: "My people will need to eat and fuel their bodies. They will have hunger to alert them to eat and I will give them nerve sensations to tell them when they are full. I want them to experience joy when they eat, I'll give them taste buds—9000 of them![1] I will engineer a tongue to secure these taste buds and wire it to their brain so that they can distinguish flavor—loads and loads of them! This tongue will also be used to formulate words, which they will use to communicate with one another and with me."

"I will let them be truly creative in having access to a wide variety of plant species. Along with a tongue, I will give them a nose, which has an astounding ability to differentiate between spoiled and fresh foods. The defined pleasure receptors should react positively to the aroma of exquisite foods."

"The edible plants need to be tasty on this new tongue, and the nonedible, not so. Three ounces of saliva per minute will help deliver the food and initiate the digestive process. The fruit should be colorful, so my people will know when the fruit is ready to eat. When they are done with the skins and cores they can be returned to the earth and replenish the ground."

God thought about the needs of our bodies in depth:

"The body will need vitamin C and potassium along with a little fiber and natural water. The banana and orange will be perfect for this. Now to engineer the trees that will have ability to bear this kind of fruit, I'll create blossoms to attract the bees for pollination and to tell them fruit is on its way! I need something to flavor up all the wheat and grains they will need. A tomato should suffice. They'll be able to slice, dice and boil them down to place on future wheat and potato sticks (noodles and french fries)."

He designed a diverse cornucopia of foods for us to choose from:

"Watermelon, pumpkins, squash, cucumbers, beans, peppers, radishes, lettuce, spinach, cabbage, garlic, celery, strawberries, raspberries, blueberries, avocados, pears, plums, cherries, and of course, the apple, pineapple, coconut, lemons and limes . . . I'll make these too. The huge variety will make eating an exciting experience and give them a variety of nutrients to keep them healthy."

"The people will need help in seeing that there is more to life than meets the eye and will discover that *invisible* forces are indeed at work in creation.[2] When they develop the technology to see my future surprises: the molecule and the subatomic, it will inspire them to look to Me in childlike wonder, and offer a thank you. It will compel them to discover more about Me!"

> God was very thoughtful to design these foods for us to fully savor, in and out of the Garden of Eden!

Each of these aforementioned garden-grown sumptuous wonders has a design story, object and purpose. All these have been initiated into existence[3] to be thoroughly enjoyed by us![4] God was very thoughtful to design these foods for us to fully savor, in and out of the Garden of Eden!

Weeds are the gardener's enemy and teacher. The good Lord wanted us to experience this first hand for several reasons; one of which was to help us to better understand certain aspects of the bible.

> A message gets across much clearer when your hands get dirty, your body becomes exhausted and your spirit is challenged!

For example, we read parables about weeds, farming and soil. Then, we learned about the challenges (or post-fall curses, thanks to Adam and Eve) in working our land and toiling by the sweat of our brow, just as God had promised to Adam in Genesis. We learned that reading and doing are very different things. A message gets across much clearer when your hands get dirty, your body becomes exhausted and your spirit is challenged! Being close to the earth can undoubtedly be very enlightening.

> The harm is not above the surface, but below it, where you can't see until you go weeding.

The foxtail looks like a harmless blade of grass with a bushy tail on its end. It is easy to think that the foxtail is not causing a problem. However, the harm is not above the surface, but below it, where you can't see until you go weeding. When you pull on the stem, a basketball-sized root system comes out! Next to a tomato plant root system, this secret invader will "choke" out your precious plant, if not kill it off completely.

> **Do not let the worries of the world choke you like a weed!**

This is a good example of how God is "talking" to us through the garden! He states in His word:

> *... but the worries of this life, the deceitfulness of wealth and the desires for other things come in and choke the word, making it unfruitful.*
>
> Mark 4:19

Do not let the worries of the world choke you like a weed! However, the foxtail was not the only plant-threatening villain to visit us.

Another weed that causes horror is Canada thistle. This weed grows fast and is very hearty. Its' needled stems and leaves make them untouchable with a bare hand, and it plagued our property with a vengeance. We talked with farmers, extension agents, organic-growers, and no one had an easy answer for how to rid these from our fields. And we had *hundreds* of them! An Amish farmer suggested brushing on a special natural chemical onto each leaf of the weed! Another grower said to burn them! What's nasty about these weeds is that they have a horizontal root system that will spread the plants 12–18 feet away! They spread like wildfire! They made weeding very difficult, choked out the good

> **You literally have to declare war to defend your crops!**

plants, they were inedible to livestock and they couldn't be touched by bare hands without penalty!

Just when you think you have gotten the best of them, the new shoots emerge. You literally have to declare war to defend your crops!

> Evil is unleashed, and if you are going to win, you can never give up!

The spiritual lesson that we gather from this is that evil is unleashed, and if you are going to win, you can never give up! This is the recipe for a fruitful garden and this is the necessary strategy to beat earth's enemy! The ground is full of enemies that would kill the fruits of your labor and steal their precious resources; the sun, the nutrients from the soil and even take their place in your field, all the while causing you heartache and pain. Is the devil, which prowls around like a ravenous lion as he waits to devour us, any different? But we are not alone in our struggles. In fact, God gives us His word that He will never leave us or forsake us.

> Design requirements of plant molecules include: formability, a memory bank to hold growing instructions, a mini-nutrient and flavor lab, sun reactivity, water-drawing capabilities and even a seasonal behavior clock.

Care and Provision

Have you ever witnessed the astounding miracle of plant growth and photosynthesis? Without the oxygen that plant life supplies, the human race could not survive. The realization of all the incredible things that God does every single day in a garden should cause us to be filled with amazement!

We need to break out the test tubes to truly consider plant functions. God needed to create life-giving molecules. God engineers DNA, atoms and ions, for starters. Some of the design requirements of the molecules include: formability, a memory bank to hold growing instructions (like the ins and outs of pollination), a mini-nutrient and flavor lab, sun reactivity, water-drawing capabilities, and even a seasonal behavior clock. All this design and engineering of growth instruction needed to be in a molecule less than the size of a pinpoint!

Two fascinating examples of this fantastic designing are the watermelon and pumpkin. From something so relatively small, these two giants develop, and epitomize the flavors of summer or fall (respectively) to a T! And all of this was created to make our taste buds happy and to build up our bodies with nutrients and vitamins.

Not only has God provided for us through the garden, He cares and provides for the garden itself. Think about the vines of these creations. These half-inch in diameter wonders carry everything the fruit needs to get a jumpstart on life. The blossoms are astounding, in that they somehow have an internal time clock, which causes them to make use of the morning dew and to pollinate. In a neighboring city to where I live, there is a business called Rockwell. They employ electrical engineers, mechanical engineers and computer scientists to program manufacturing equipment. These specialists program the machines. The programs instruct the machines to perform a variety of tasks. These tasks include: what temperature, how long to hold the temperature, when to change the temperature, and when to end the task. The capabilities of these *man-made* machines are fantastic!

> God has programmed a "command center" into all living things.

Like these engineers and scientists, God has a programmed a "command center" into all living things. The squash family has a yellow die injection-system for the plant's blossoms. These bright blossoms attract the bees that pollinate them. The blossom leaf is so delicate that a strong wind will tear them. They seem to open and close as if on cue. There are both male and female flowers. Each has their specific job.

Because of their fragility and fertility make-up, the female flowers need protection from predators and the heat of the sun in their early weeks of development. A dozen flowers may emerge, but only a few will develop fruit. The thin vines are engineered to produce giant, umbrella-like leaves to cover shorter, fragile female flowers. How can there be any question as to the deep forethought required to produce such intricate vegetation mechanics?

> How can there be any question as to the deep forethought required to produce such intricate vegetation mechanics?

The male flower is engineered to grow higher than the one-foot leaf canopy. His pollen-coated anther attracts the bees from afar, but this also attracts the deadly cucumber beetle. This design can cause enough of a distraction above the canopy to allow the discrete pollination process to happen under cover.

If fruit has been produced, the baby pumpkin or watermelon will need lots of water. The wonder vine (which is more like a tiny stick at this stage) gets a little bigger, but can't quite keep up with the water demands, so out pop tendrils (the cute curly-cue things that decorated Cinderella's pumpkin carriage). These are actually water augmenting, ground-boring vine tendrils (I think of them as God's de-*vines*) and they are hard at work. These spinney little guys drill into the ground and draw water to the vine. I'd like to see any human being attempt to engineer and design all that on a pinhead!

Sweat of the Brow

We can see that God did his part in the fantastic and glorious design engineering of these plants—and all vegetation, as far as that goes. In the meantime, mankind is still trying to find the control box in the vine that initiates all the perfectly timed creation commands!

We have been created with an innate drive to protect the source of our sustenance. The sweating begins with the fear of unknowns.

The Garden Around Us   103

So begins the anxiety of not knowing what may lurk in the fields of our labor and our quest to conquer anything that threatens to destroy it. Other unknowns include airborne micro-bombs like blight and visible plagues like powdery mildew. All the while, the life-lesson-analogies begin to take root—a well designed plot is in the making.

People who are not farmers think that once the plant is growing the work is done! Now all you have to do is water it and sit back and wait until harvest time! Not!

Soon, the dreaded invaders appear, like the Hornworm. It looks like some kind of strange, alien, five-inch long, green caterpillar Dachshund mix, with a tail that doubles as a stinger! In a matter of days, they can decimate an entire crop. Some may assume deer were just enjoying the produce as an evening salad bar, but later learn that it was this insidious caterpillar! The Hornworm comes prepared for his secret mission of destruction, completely camouflaged with the exact color of a tomato plant leaf. To find them, just listen for the crunch of their jaws slicing through the tender plants as they drive across the field. The sound is unmistakable.

Another example of how the farmer toils, as predicted in the book of Genesis, is blight. Blight is a disease which occurs if the moisture is just a little too high and the temperature drops too low. Blight is another invisible invader doing its evil deed. It will wipe out fields of tomato plants in less than five days if not treated.

We may think the sweat of the brow ends with plowing and planting. Ha! Now the farmer must remedy the situation quickly, or the plants will fail.

> **Mankind is still trying to find the control box in the vine that initiates all the perfectly timed creation commands!**

Remedies may include: plucking, spraying, wiping, squishing, and, worst off all, doing all of these tasks while bouncing around on a tractor, or stooped over in the heat of the sun. In fact, many times the only way to really eliminate the threat is crawling around in the dirt on our knees. This toil will help us to appreciate Heaven all the more!

In the present days of modern conveniences, it is easy to forget that our ancestors were forced to deal with how quickly produce can perish. The first thing on their minds, we can be sure of, was *not* to put the newfound veggies in a two-door Frigidaire! They quickly discovered another use for the cave with its cool temperatures.

> **This toil will help us to appreciate Heaven all the more!**

Once the days of harvest arrive, the farmer's brow is again covered in perspiration. He will rush to enroll in the local farmer's markets to sell his produce before it goes bad. He will have to prepare for the market by purchasing scales, have the scales audited, buy tables, signs, umbrellas, etc. If he doesn't own a truck, he will ask friends and neighbors to lend him one. He will need to buy containers, pick the produce, pack the containers, price them, set up his booth, talk to customers, and encourage them to buy his produce. Usually, by noon or so, most farmers take down their display, unpack, dispose of the "unsellables," and repeat the entire process again for each market. We should never question the price of a tomato again from a local farmer! His *sweat of the brow* is done on our behalf! He is managing the garden around us! Thank you farmers! Maybe the real roots of celebrating Thanksgiving is that the farmers are thankful that the season is finally over! And we give a *healthy* "thank you" to our Master Gardner Creator that we have the ability to keep our bellies full with just the right ingredients!

> *"Gratitude is one of the greatest Christian graces; ingratitude, one of the most vicious sins."*
>
> Billy Graham, American Clergyman

Perfection

It's all in motion, coming and going, preparing, plowing and planting. All the while, the plants are sprouting from a seed, then producing fruit and, finally, decaying. Each garden element at any given time is in a state of change. All designed *intentionally* for necessary life-sustaining order and energy for our bodies! Along the way, I am sure that God wants His children to enjoy all that He has made: A garden so perfect that it involves a three-part symbiosis that joins man, seed, and God—as perfect in function then as it is now (in spite of the challenges the farmer faces).

> God designed this garden for *us*; to capture our imagination and to prompt us to wonder about the invisibles of life.

God designed this garden for *us*; to capture our imagination, and to prompt us to wonder about the invisibles of life. He wets our palate with the promise of a curse-free garden to come. In reading about and experiencing this garden, we can draw analogies for shepherding, guiding and leading. We can gain new insights on discipline and perspective and enjoy physical, mental and spiritual health. In God's garden, there is rest, reflection and restoration.

Before the beginning, extremely complex ideas and concepts, with much consideration to heart related passions were on the drawing board of life. This divine mix of mass-energy equivalence formulas and directives interplaying with thousands, if not millions, of sub-atomic based clusters lie in wait for life emergence. The emergence of the tangible will connect built-in information streams designed to deliver time-sensitive, life-giving commands. These

> Before the beginning, extremely complex ideas and concepts, with much consideration to heart related passions, were on the drawing board of life.

life-receiving creations stand ready to engage the photosynthesis process. All of this was designed and engineered with perfection, before the beginning, *simply waiting on the breath of life to come.*

God's handiwork and His loving care in design are indeed something very special to think about. Our worship would rise to Heaven's gates if we would simply let the shouts of creation echo in our hearts.

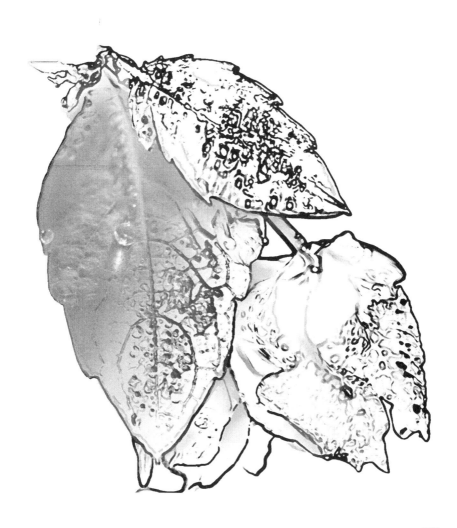

7

1.2 oz–2,000 lb Beasts

Our Creator went to a lot of trouble so we can enjoy a good steak, burger, roast and corned beef sandwich! Do you see the creativity of the bovine in the soccer balls, baseball gloves and leather jackets we might use in everyday life? Many of today's industries are built around this noble beast—the cow. Perhaps McDonalds® [1] wouldn't be who they are today if they were based on a hog or a tuna fish. And when you really explore all of the uses this multi-functioning creature possesses, once again, we should experience some shock and awe! How many people reach for cream in the morning for their cup of tea or coffee? Milk with chocolate—a symbiosis bar none (pun intended)! Shakes, cakes, puddings and cream pies are all cow dependant!

> Shakes, cakes, puddings and cream pies are all cow dependant!

Where would the salad, or cracker or chip be without cheese? We dip with it; we smother with it, age it, smoke it, grate it and shake it. It's a flavor that tantalizes our taste buds!

How many of us can down a quart of ice cream in a matter of minutes? TCBY®, Dairy Queen® and Ben and Jerry® [2] would be lost without our friend the cow! Lone Star® would be lonely without this hoofed wonder. Outback® [3] would be held back without the Black Angus beauties.

Nothing else converts green grass into ground-round like a cow! Cow basics include a flyswatter tail, four-stomach, cud-chewing digestive system, a 13.5 gallon daily water-injection system (saliva), a highly random manure spreading system to help fertilize the fields, and a growth rate of 60 to 1,200 lbs or more in 18 months just on grass alone!

Like our pets, cows like a treat. In this case—sweet feed! I thank God for this. *Because* cows naturally like to eat the grass on the other side of the fence, they sometimes escape. When they escape, it can create dangerous situations, but God gave us the ability to tame and control these massive creatures! When cows see a feed bucket, they come walking (or sometimes running), which makes it relatively easy to lead them back to pasture or back to the barn for the night. I have actually seen cows that are so happy that they frolic, like a puppy that was shown a stick. It is little nerve-racking, having a 1,000 lb creature happily kicking up its hooves as it races toward you.

> Nothing else converts green grass into ground-round like a cow!

Another instinct of design is that cows seem to want to obey. In raising them, I had the personal experience to see them obey a "stop" hand signal in the middle of a stampede (I am thankful I did not die in the midst of such a demonstration). The cows did this without attending a single cow obedience class! God made them to pay attention to us. When they don't pay attention, there are some dog breeds that know how to get them to obey. God made the cows "corral capable"!

In fact, The Creator made many creatures that defy physics and biology in a single bound. These living things cannot function or exist in a phase of its' own evolution. By examining how these creatures operate, it is easy to see that they were designed to function just the way that they do today.

Animals That Challenge Us!

Giant Sea Boogie Monster. It's not a submersible flying carpet, or a billowing "blob-aceous," or an abandoned parachute that is inhabited by a giant snail. Steven Haddock, a scientist for the Monterey Bay Aquarium Research Institute in Moss Landing, CA, says that the mysterious creature is a deepstaria enigmaticajellyfish. It's translucent, has a honeycomb-like internal tissue or cell structure, can turn itself inside out, and has an arm-like organ that comes in handy (pun intended). A recent video shot, dated April 25, 2012, reveals the mysterious creature

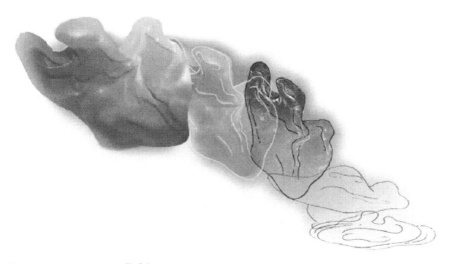

Deepstaria enigmatica jellyfish

110   Before the Beginning

of the deep, resting at about 5,000 feet down next to an oil well pipe. This particular jellyfish was captured on video by Oceanic. Naturally curious *deepstaria enigmatica* jellyfish hardly seems to function under its' own command. It looks like a ghost-like hooded transformer of sorts, with very few parts. If it could not undulate so magnificently, it would likely sink to the bottom and die. The graceful movements are vital to its existence.

Some other intriguing jellyfish features include: an internal electricity production and storage facility, an erythematic light pulse transformer and a self-defense electrical discharge taser unit, which can stun or kill upon contact! Although the purpose of this jellyfish has yet to be discovered, the *enigmatica* is full of mystery. God created these just because He could! Perhaps God wanted to send a challenge our way and wanted us to explore to find out more. Search this video out online and be amazed.

Mimic Octopus. The name says it all! It is said that it can mimic at least 15 other species of animals. The built-in knowledge, along

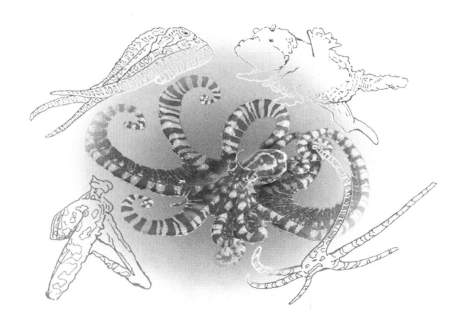

with an extremely clever instinct, is what makes this eight-legged sea serpent something to reflect on. The anatomical features and responsive molecular structure designed into this *one* creature is astounding! Think about it: The ocean does not come with full-size mirrors to practice impressions! In a recent interview, the comedian Jim Carrey said that many of his early years were spent in front of a mirror contorting his face into celebrities such as James Cagney or Clint Eastwood, or just exploring "where no man has gone before" with his face. He would spend hours in front of a mirror getting the morph and mission just right.[4] The mimic octopus gets its impersonations right the first time without practice in the mirror. Some of the animals it mimics include; the flounder, lionfish, sea snake, stingray, and it can even transform itself to look like a turkey with human legs! It can also change color instantly to match the environment and it squirts the usual octopi dye (a necessary predator protection camouflage).

Dana Carvey, in the humorous family friendly movie, *Master of Disguise*, played the part of a human chameleon. He changed from turtle guy, to cherry pie, to cow pie without missing a beat! He did this with the full capacities of being human (with the help of a prop and a costume design shop). Our mimic octopus needed to be complete in design from the moment it appeared in the ocean in order to pull off a performance such as this. If it failed in a half-developed presentation, it would simply be eaten. At

> The mimic octopus gets its impersonations right the first time without practice in the mirror.

that point, it could not genetically pass on its "learn how not to be eaten" experience. I can't help but see God designing the mimic octopus and having a lot of fun with it! One might even have been able to hear a zany movie quote coming from God's Creation lab, "It's so crazy . . . it just might work!" So next time you are digging into your octopus calamari and get one of those tentacles sticking to the roof of your mouth, remember the mimic octopus and it's wondrous Creator!

112   Before the Beginning

Bombardier Beetle. This insect reveals God's nano technology at its best! The bombardier beetle is a miniature traveling chemical mixing plant and army tank all wrapped up in one tough little package! This half-inch bug makes up its own hydrogen peroxide and hydroquinones to make smoking, caustic benzoquinones. This chemical explodes out at 212° F through rotating turret style twin nozzles with precise bull's eye accuracy on demand. This design offers a safe containment system along with an inhibitor—another control chemical. If this bug were half-developed through an evolutionary process, we would have a containment breech

> One might even have been able to hear a zany movie quote coming from God's Creation lab, "It's so crazy ... it just might work!"
>
> Pistachio Disguisey
> in the movie
> *The Master of Disguise*

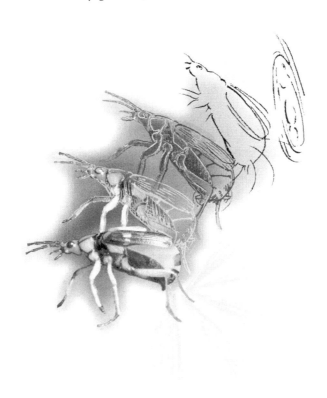

and the bug would dissolve itself![5] This defense system would need to be securely intact with all systems "hot" to function properly. This little bug blows away any of man's understandings through its highly sophisticated engineering, while pointing its inception to an extreme Designer.

Volcano Blind Shrimp. Geologist Bramley Murton from the RRS James Cook states:[6]

> "On the ocean floor at a volcano opening, temperatures can reach 750 degrees F, heating water to the point where it can melt lead. The blazing hot, mineral-rich water is expelled into the icy cold of the deep ocean, creating a smoke-like effect and leaving behind towering chimneys of metal ore, some two story's tall. The spectacular pressure—500 times stronger than the earth's atmosphere—keeps the water from boiling."

But Murton points out creatures that survive in spite of impossible circumstances:

> "The environment may appear brutal: the intense heat and pressure combines with toxic metals to form a highly acidic undersea cocktail. But vents host lush colonies of exotic animals such as hairy worms, blind shrimp and giant white crabs."

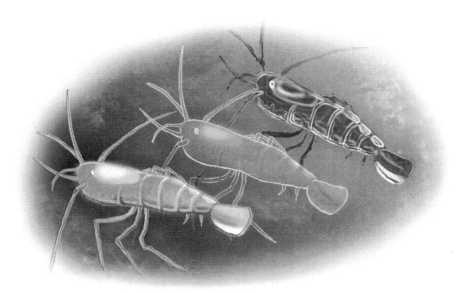

Murton states that the environment appears "brutal." While this habitat may serve these particular beings quite well, for us, describing this environment as brutal is an understatement. These temperatures alone melt steel! In these conditions, the tissue of any other animal would become completely annihilated. Yet, these animals are thriving. This is beyond physics or biology as we know it! As God spoke to Job to adjust his perspective, likewise God may be adjusting our perspective by exhibiting divine genius in His creation.

> *For since the creation of the world, God's invisible qualities—his eternal power and divine nature—have been clearly seen, being understood from what has been made, so that men are without excuse. For although they knew God, they neither glorified him as God nor gave thanks to him, and their foolish hearts were darkened.*
>
> Romans 1: 20–21

The Angler Fish. Fishing is a way of life for many and it can mean many things to many people; from pleasure to employment, but for the angler

fish, it means survival. This is one of the few light-producing animals on the planet. It has a very practical tool for attracting prey. This fish was designed to live about a mile down on the ocean floor, in complete darkness. There is a light on the end of a narrow extension that is attached to the top if its head. Others may say it's more suited for an intimate, candlelit dinner for two! The only problem with that idea is that the guest just happens to be the main course! The pole also has a bait formation at the end of it. The light leads the way and attracts its prey.

Another major difference between angler fish and other fish is that it doesn't have buoyancy control. No swim bladder specifications can be found in the blueprints of this fish. The prominent bait and pole displayed in the shallows would make the angler fish an easy target and it would likely not survive if it had a swim bladder. The blueprint is complete and accurate just as God intended.[7]

The European Green Woodpecker. When this neighborhood exterminator makes his daily feeding rounds, he is actually keeping

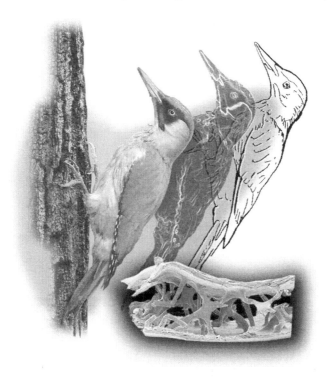

our trees healthy. God demonstrates His love for us from the tip of this bird's tree-grabbing toes to its cute little drumming nose. And the beak is unlike any other beak in strength! If other birds had to hammer like this

> As God spoke to Job to adjust his perspective, likewise God may be adjusting our perspective by exhibiting divine genius in His creation.

guy, the results would be less favorable. A canary's beak would collapse. A chickadee beak would be chiseled away. A sparrow's beak would be splintered. But when purpose is considered in the creation design process, God understood that there was a need for a very strong, precision-built, 5-in-1-bug retrieval tool. This mechanism had multiple purposes: it had to be able to knock on wood, reach into the tiny holes it drilled and grab insects and grubs and then, using that same beak, it had to be able to breathe! Another design requirement was the tongue. It's long tongue retraction system recoils around the head from top to bottom. The green woodpecker's tongue has barbs on the tip to drag its prey back through very small tunnels, all the while excreting special glue that holds the prey, but avoids gluing the bird's mouth shut! God designed the woodpecker's head to be protected from constant impact by including a special impact absorbing cartilage as part of the structure (quite a demonstration of mercy for this little creature). And for fun, God adds brilliant green coloring in the feathers for decoration. The bird has an incredible migration system, feather stabilizers for tree perching and an extra toe on the back part of the foot for clutching tree branches! All but two of the woodpecker species have the extra toe.[8]

> God understood that there was a need for a very strong, precision-built, 5-in-1 bug retrieval tool.

Australian Incubator Bird (Malleefowl). This particular species of ground dwelling bird looks sort of like a 3.5–4 lb. turkey. What's so unique about this bird is its uncanny ability to regulate temperature and humidity within its nest. The nest size is unbelievably large—about 20 feet across and has been seen to be as high as 50 feet! It also goes down about three feet into the earth. The nest is made and managed by the male. Sometimes multiple nests are made if the female isn't quite satisfied with the first construction the poor fellow has assembled. The Malleefowl dedicates nine months to a year building and maintaining this large incubation mound of soil, leaves and twigs. The Malleefowl maintains a mound temperature of 91° to 99.5° F and controls humidity by using its beak as a thermometer and hygrometer. They also can adjust the soil cover to either retain or expel heat from the egg chamber by digging out soil or adding it.

The egg is almost the size of an ostrich egg. This comes from a four pound bird! When the egg hatches, the chick lies on its back and starts to scratch at the ceiling. Debris falls on its chest. Undaunted, it simply gives a shake and scratches again and repeats this escape process, which continues for up to three days. Talk about perseverance!

The survival rate with the Malleefowl is less than 2%. All this effort to break free from the confines of the eggshell, and in the next phase of its life, there is little chance of survival due to predators! According to the Malleefowl Preservation Group, this rare Australian bird is currently threatened by extinction.[9]

This story may mean different things to different people. We should be in awe just thinking about this amazing beak. How can this boney material measure temperature and humidity? This bird is designed with extremely accurate sensors that are "wired" to the brain, and the bird responds faithfully, time and time again. We humans can study the Malleefowl and discover usable scientific information, but if we look at this bird through the eyes of our Creator, we see another reason for its existence. Perhaps this bird was created to declare in a loud voice that no man can know the mysteries of God! It stands as a testimony to absolute (not accidental) *perfection* in design. For example, if the baby bird were to start scratching down like a chicken, it would never make it out of the nest and would ultimately die. It instinctually knows in which direction to dig and it knows not to give up!

The design and foundational work of these animals is complete and astonishing.

You can't use half-created or partially evolved raw elements to create something. These common elemental atoms need to be fully intact for them to have the complete behavioral properties. Half developed carbon, nitrogen, phosphorus, oxygen, sulfur and half developed neurons or atoms with a half-baked strong force wouldn't do it either. These and other raw elements need to be fully complete to function. Same as a fully developed animal, if they are not complete to begin with, they would not survive. If not fully functional, the "escape from predators" mechanisms that are in a complete animal design would not permit an escape; they would be eaten! Hence, an evolution or change, could not take place.

The passing of time, in and of itself cannot create. All matter and energy breaks down over time. Organic things always rot. An example is if an attic full of junk sat for a million years, it would turn to a pile of dust. To believe that a higher ordered form will eventually emerge from the piles of dust is what the core of evolution is all about. From the deceiver's point of view, it's somewhat comical or pathetic, I don't know which, that the devil is attempting to mock God. If he were to comment on this it would sound like: "Look what your creations believe in! They are so easily deceived. I can get them to believe in the ludicrous—instead of understanding deterioration, I will get them to believe that deterioration 'over time' will result in perfect-ordered creation. Watch, I will use *their self-perceived strengths of intelligence* to blind, not only themselves but others as well!" The common phrase: "You can't see the forest through the trees" couldn't ring truer.

The message for us is—perseverance! Could God be challenging our prideful, self-sufficient attitudes? Could He have put these mysterious, mind-boggling animals in our midst to utterly blow away our traditional, finite thinking minds?

Pride is difficult to recognize, let alone overcome. Creation brings many of us to our senses. It drives us to consider and acknowledge God, and can force us to humbly get out of the way and recognize that we are the children of an amazing, indescribable Creator God who loves us with a passion beyond measure!

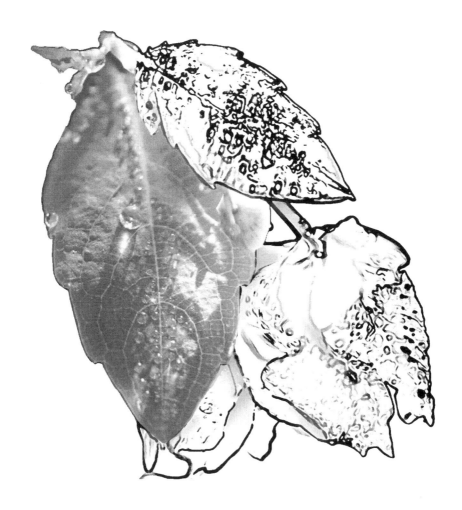

8
The Masters and Masterpieces

A masterpiece is something that was created, developed or discovered by someone, which others have established as rare and of high value. In many areas of life, we have granted individuals the title of being a "master." The arts, sciences, geology, physics, painting, music, drama—even a sharp shooter—can reveal the skill level of various "masters." The question we must ask ourselves is:

Were the earthly "masters" and "geniuses" of history self-made creators and thinkers or were they divinely inspired?

Jesus stated that we would be doing greater works than these:

> *I tell you the truth, anyone who has faith in me will do what I have been doing. He will do even greater things than these, because I am going to the Father.*
>
> John 14: 12

Maybe its not one or the other, maybe it's a little of both!

In *The Book of Genius*, Tony Buzan and Raymond Keene made the world's first attempt to rank the top ten geniuses of history. They included:

10. Albert Einstein
9. Phidias (architect of Athens)
8. Alexander the Great
7. Thomas Jefferson
6. Sir Isaac Newton
5. Michelangelo
4. Johann Wolfgang von Goethe
3. The Great Pyramid Builders
2. William Shakespeare
1. Leonardo Da Vinci

Each of these "greats" has created awe-inspiring work. When created works deviate from what we know or that which is familiar, we consider that it was divinely inspired and has the fingerprint of THE master. Through the works of these artists, God makes Himself evident. What about the rest of us less inspired people? Even J.K. Rowling, the author of the very successful *Harry Potter* series, shared that what troubled her most was that she felt she was an impostor—just a regular person—and not the genius that the world was making her out to be.[1]

When I was working on a design project for a large lighting manufacturer, I remember calling the inventor of some innovative 3D software. One of the first commercial users of this software was a very well known animated movie production company. In fact, the software was so successful that one of the demo projects is still being used as part of their logo today. When the software came out, it boasted texture mapping and included recognition of light source and shadow. These software features give the graphics a photo realistic appearance. The software even has an organic plant-growth mode! You set some end-result parameters and the software generates virtual growing branches, stems and leaves! As an artist, this new capability blew my understanding of the norm. Drafting boards would soon become obsolete, and it was the beginning of the end of hand painted cell animation.

Questions filled my mind. Who can possibly think like this? Is he real? With this type of creative ability, I had to meet him! After just a few phone calls, we were able to connect. The first thing I asked the inventor was how he could create something so fantastic? He was very kind and modest. He said, "It was no big deal, just simple math equations." Yeah, easy for him, but *I* have never had a clue on where to begin the designing of such a tool! I realized that this would change the playing field entirely. And it has! Today, at least a third of the movies that are produced are fully digitally created, and a majority of them use this inventive technology for special effects. Whenever and wherever visualization needs to happen, you'll find the 3D digital technology of this "master." I can't help to wonder if software creation can be divinely inspired! If it isn't, then developers are certainly using

> When created works deviate from what we know or that which is familiar, we consider that it was divinely inspired and has the fingerprint of THE master.

their God-engineered brain to design this spectacular technology. Either way God is glorified!

> When you talk to most creators, whatever they did, they will say it just came naturally to them.

When you talk to most creators, whatever they did, they will say it just came naturally to them. They don't feel special; they put their pants on one leg at a time just like the rest of us. But Mozart was composing when he was only five years of age! It had to be natural for him; he heard it in his head and wrote it with his heart! God fully knew the gifts that Mozart possessed from the moment he was being formed in the womb—even before that! The musical beauty of this "master" is nothing short of astounding! A natural implication one could draw is that God wanted to demonstrate to us that life is bigger than the limits we have placed upon it. In Mozart, God used a *child* to change the world of music and to reflect His glory.

In life, we attempt to make sense of the world around us and we create small "boxes of understanding." But God doesn't want us living out of these boxes. Life is grand and through God's unmerited favor, He lets us experience His magnificence! God can use whomever He pleases.

In his book, *How To Think Like Leonardo da Vinci*,[2] Michael J. Gelb focuses on one of the world's most recognized geniuses. Without a doubt, the entire world considers Da Vinci to be the epitome of the term "master." Some of Da Vinci's best known works include: the fresco painting of The Last Supper, the Mona Lisa, the Virgin of the Rocks and Madonna and Child.

> God wanted to demonstrate to us that life is bigger than the limits we have placed upon it. In Mozart, God used a *child* to change the world of music and to reflect His glory.

Da Vinci was also a renowned architect and sculptor. His understanding in the areas of anatomy, botany, geology and physics were reflected in his studies, drawings and notes. He put many of these disciplines into use in his sketches of modern flight, helicopter, parachute, extendable ladder, armored tank, machinegun, mortar, guided missile and submarine, much of which is even used today. He pioneered automation, invented the three-speed gearshift, a thread cutting machine for screws, the bicycle, monkey wrench, snorkel, hydraulic jacks, waterwheel, folding furniture and more!

> Without a doubt, the entire world considers Da Vinci to be the epitome of the term "master."

Gelb wanted to find commonalities such "geniuses" shared in their approach to life and learning. Listed below are the seven observation principals that Gelb discovered in his research:

Curiosita: An insatiably curious approach to life and an unrelenting quest for continued learning.

Dimostrazione: A commitment to test knowledge through experience, persistence and a willingness to learn from mistakes.

Sensazione: The continual reinforcement of the senses (especially sight) as the means to liven the experience.

Sfumato: Literally meaning "going up in smoke." This refers to a willingness to embrace ambiguity, paradox and uncertainty.

Arte/Scienza: The development of the balance between science and art, logic and imagination; "whole-brain" thinking.

Corporalita: The cultivation of grace, ambidexterity, fitness and poise.

Connessione: The recognition of and appreciation for the interconnectedness of all things and phenomena; "systems thinking."

Gelb's book goes into more detail on each of these principals.³

I would gather that each of the aforementioned greats would excel a little more or less in each of Gelb's principals.

The Holy Spirit helps us to ask the right "what if" and "why" questions.

> *And I will ask the Father, and he will give you another advocate to help you and be with you forever.*
>
> John 14: 16

> **The Holy Spirit helps us to ask the right "what if" and "why" questions.**

I would ascertain that we would all do well to carefully listen to the Holy Spirit's quiet whispers within. If we can work on dropping the unchangeable past and focus on the unwritten future,⁴ the breeding grounds of invention and discovery would be tilled.

> **Our sinful nature can negatively affect our ability to create.**

Our sinful nature can negatively affect our ability to create.⁵ There is evil running rampant, manifesting itself as greed, envy, lying, cheating, stealing, bearing false witness, lust, adultery and the list goes on. Now imagine life for a moment without these influences and tendencies. Picture God working with us in a sinless state; imagine a "pure" humanity, with a clean countenance and conscience.⁶ Without these distractions, the creative process wouldn't be hindered!⁷ We indeed would be unencumbered masterpieces of God, reflecting more and more the image of *the* Master.⁸

Our approach to seeing God in the works of His people should open us to new revelations. God is about mystery, which is what makes life an adventure. These works don't have to be world-renowned to be legendary! The pursuit has to be instinctual, almost inhibition free; not

The Masters and Masterpieces   127

caring about what anyone would be thinking of you in your pursuit, like a child grabbing for a prize! The last time I saw this look in a child's eyes is when I put on a "dragon training" themed birthday party for my son. I had rigged up a piñata-shaped dragon on a low-flying zip line and gave each child a turn with a wooden sword. The fight and determination I saw in their faces was remarkable! Each boy stepped forward, eyes bulging and tongues swirling, as they swung their swords to slay the dragon that was flying right into their faces. Their fears were conquered! Like these boys, I would like to be completely unencumbered and absolutely passionate in all my pursuits, including exploring the wonders of God and the mysteries of His creation.

The boys became one with the their sword and their mission. A mission prewired in their DNA, which could swim around on the surface the size of a gnat's knee!

Richard Dawkins, the author of *The God Delusion*, is described as the world's most famous atheist and evolutionary biologist. Dawkins appeared at the Sheldonian Theatre in Oxford, England on February 23, 2012. His dialog shocked many listeners who attended the debate, which was titled, "The Nature of Human Beings and the Question of their Ultimate Origin." In spite of being a life-long proponent of maintaining a scientific perspective, Dawkins stated that he could in no way disprove God's existence. Based on that, Dawkins admitted that he could not be an atheist since he could not prove that God does not exist. This stance moves him into the realm agnosticism. Some may put Dawkins' early works in the "genius" category. One can be considered a genius and still change one's position regarding the existence of God. Perhaps in a later edition of this book, I will move Richard Dawkins to the "Transformers" chapter if he makes the complete journey from atheism to believer.

> **I would like to be completely unencumbered and absolutely passionate in all my pursuits.**

Here is a list of history's famous scientists who believed in God:

1. Nicholas Copernicus (1473–1543)
2. Sir Francis Bacon (1561–1627)
3. Johannes Kepler (1571–1630)
4. Galileo Galilei (1564–1642)
5. Rene Descartes (1596–1650)
6. Isaac Newton (1642–1727)
7. Robert Boyle (1791–1867)
8. Michael Faraday (1791–1867)
9. Gregor Mendel (1822–1884)
10. William Thomson Kelvin (1824–1907)
11. Max Planck (1858–1947)
12. Albert Einstein (1879–1955)

Einstein and Isaac Newton made it to the top ten geniuses list as well! Einstein is probably the best known and most highly revered scientist of the twentieth century and is associated with major revelations in our thinking about time, gravity and the conversion of matter to energy ($E=mc^2$). Although never coming to belief in a personal God, he recognized the impossibility of a non-created universe. The *Encyclopedia Britannica* says of him: "Firmly denying atheism, Einstein expressed a belief in a . . . God who reveals himself in the harmony of what exists."

> However, one can be considered a genius and still change one's position regarding the existence of God.

This actually motivated his interest in science, as he once remarked to a young physicist: "I want to know how God created this world; I am not interested in this or that phenomenon, in the spectrum of this or that element. I want to know His thoughts—the rest are mere details." Einstein's famous epithet on the "uncertainty principle" was "God does not play dice"—and to him this was a real statement about a God in whom he believed. A famous saying of his was "Science without religion is lame, religion without science is blind."[9]

Sir Isaac Newton was an English physicist, mathematician, astronomer, natural philosopher, alchemist and theologian, who has been "considered by many to be the greatest and most influential scientist who ever lived." Newton described universal gravitation and the three laws of motion, which dominated the scientific view of the physical universe for three centuries. Newton set standards for scientific publication down to the present time.[10] Newton built the first practical reflecting telescope and developed a theory of color. He also formulated an empirical law of cooling and studied the speed of sound. In mathematics, Newton shares the credit with Gottfried Leibniz for the development of differential and integral calculus.[11] His list of great works goes on and on.

> Which God-believing scientist will be next on the list?

Newton, although an unorthodox Christian, acknowledged God's hand in creation. Newton saw God as the masterful creator whose existence could not be denied in the face of the grandeur of all creation.[12]

Which God-believing scientist will be the next on the list?

Dr. William Lane Craig, theologian with a double doctorate, defender of historic Christianity, debated with Dr. Peter Atkins, the proclaimed atheist and renowned biologist. Dr. Craig stated that science cannot account for everything. He listed five things that cannot be scientifically proven: 1). Logic and mathematics; science presupposes logic and math, so trying to prove it by science would be arguing in circles. 2). Metaphysical truths. 3). Ethical statements of value. 4). Esthetic judgments. 5). Science itself cannot be justified because it's permeated with un-provable assumptions.

I would think that the idea that science cannot account for everything would come as a huge relief to scientists! Da Vinci learned the art of embracing that which is immeasurable. Piercing clarity would come out of the sciences if they would remain in the confines of what is measurable and include God as God in their equations.

> I would think that the idea that science cannot account for everything would come as a huge relief to scientists!

Professor Dr. Heribert Nilson, a botanist at Lund University in Sweden, concurs with this line of thinking. In his book, *Synthetische Artbildung* (The Synthetic Formation of Kinds), Nilson states, ". . . the theory of evolution is a severe obstacle for biological research. As many examples show, it actually prevents the drawing of logical conclusions from one set of experimental material. Because everything must be bent to fit this speculative theory, an exact biology cannot develop."[13]

Dr. Peter Atkins stated that it would be "lazy" to give recognition to God and it is only bored and desperate people who need to believe in Jesus Christ. As Dr. Atkins fires his shots over the bow, I would suggest that his gifted mind of analysis is incapable of analyzing and considering God. Perhaps if he allowed the Spirit to bless him, he would be able to account for the immeasurable. Believers would do well to pray for unbelievers in the science community, for this is God's way of dismantling the work of the enemy. By the way, as uncomfortable as involvement in a spiritual battle may be, the strategic commander is also the one who abounds

> In this war, there is no draft or forced enlistment.

in grace and mercy. After the dust settles, He provides opportunities for everyone to choose a side. In this war, there is no draft or forced enlistment. The Creator's placement of free will in every member of the human race is of utmost importance and enlisting on the wrong side has dire, eternal consequences.

The Masters and Masterpieces 131

The Battle of Anghiari
Peter Paul Rubens' copy of "The Battle of Anghiari"

Seek and You Will Find

On March 12, 2012 a magnificent discovery was made. The long lost Da Vinci fresco titled "The Battle of Anghiari" was discovered on a wall, behind another wall depicting Giorgio Vasari's fresco titled "The Battle of Marciano." Da Vinci's masterpiece was discovered in the Plaza Veccio, Florence Italy's 14th century City Hall. However, to get to Da Vinci's piece, the works of Vasari had to be drilled through to get to the cavity where "The Battle of Anghiari" was hidden. With probes, experts were able to carefully take samples. These samples have been analyzed with a scanning electron microscope. A black substance revealed an unusual chemical makeup of manganese and iron; this was the same black pigment compound found in brown glazes on Leonardo's "Mona Lisa" and "St John the Baptist." They have matched up Da Vinci's fingerprints, as well as the layering techniques that are found in Da Vinci's work.

The researchers drilled in previously restored areas and fissures, not disturbing the original Vasari. They also took into consideration that Vasari was protecting Da Vinci's work. Whether right or wrong, they made the determination to reach one of the biggest finds in art history in decades. In time, perhaps the mystery behind the hiding of Da Vinci's work and its discovery will be revealed.[14]

> God longs to reveal His Masterpieces!

Like those who eagerly uncovered the work of this artistic master, God longs to reveal His Masterpieces. Through His Word and the divine moving of the Holy Spirit, the very soul of our being can be discovered. God uses the Holy Spirit, like the drills and probes used to reach DaVinci's Masterpiece, to pierce the bone and marrow of our very being. Our pride, ego and even the lack of understanding serve as the layers that God has to drill through to get our attention. God's word is described as a two-edged sword, piercing bone and marrow.

Cropping of the "Battle of Marciano" painting.

What an incredible thought to ponder. In the words of Phillip Yancey, "... the universe was God's own work of Art, and the human body God's Masterpiece."[15]

When we pursue God's truth, He tells us that we will find it. While examining the Vasari painting, a tiny flag was discovered. This flag does not appear to be connected to the scene. It reads: *Cerca trova*. This phrase is Italian for "Seek and you will find."[16] Some experts believe this to be a "note" left by Vasari to indicate the hidden work behind his own.

You and I are the Masterpieces of God!

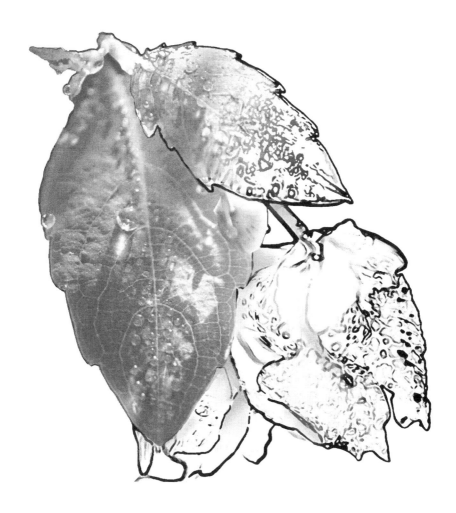

9
Unveiling Creation Design

I sometimes wonder if one attribute of God is that He is modest. God is not jamming His authorship of everything that exists down our throats. He is leaving our acceptance of the true creation origin in our own hands. He is asking us to lean on faith rather than our own understanding. As the title of this chapter implies, there can be a "veil" shielding our view. At presentation ceremonies, there is an appropriate time to "reveal" whatever it is being featured. The presenters write an agenda of items that need to be addressed prior to unveiling. This adds drama and meaning to the event. Some items that may be covered prior to the unveiling include: the process, who did what, how long it took to create it, what sacrifices were made and the accolades of all involved. Next, as the veil drops and as the item is revealed, applause and delight follow. We do this because it's fitting for the occasion. I wonder if God is conducting a ceremony of sort as His creation is unveiled to each of us. Perhaps as we began our walk with the Creator, we couldn't fully appreciate His handiwork because we were not able to see it. I have a missionary friend in Istanbul and this experience is very clear to her. When someone gets the big picture of faith, they start to see beauty for the first time. Not that this is a soul barometer by any means, but an observation found in her area of ministry. Whatever the case, as we begin to see more and more, the appreciation of beauty and His handiwork are one of the rewards we get for seeking His Kingdom!

> He is leaving our acceptance of the true creation origin in our own hands.

> I wonder if God is conducting a ceremony of sort as His creation is unveiled to each of us.

Change Blindness

National Geographic examined an intriguing human behavior on one of their programs.[1] The camera crew created several situations, similar to the "Candid Camera" style of years gone by, to conduct some experiments. One scenario was done with a woman on a blind date. She was so focused on the male-female interaction that she was completely blind to the extensive activity

> When someone gets the big picture of faith, they start to see beauty for the first time.

going on around her. While the woman was engaged conversationally, the crew changed the entire restaurant wall color, themes, furniture and fabrics from a Chinese restaurant to a Mexican restaurant! She didn't see a thing! The term for this "limited view" is called *Change Blindness*. It refers to a blindness (or veiling) that occurs when our attention is focused so intently on one thing that we neglect to notice everything else around us.

> Blindness (or veiling) occurs when our attention is focused so heavily on one thing that we neglect to notice everything else around us.

The National Geographic team set up another scenario with a hotel clerk registering hotel guests. Upon the arrival of each new guest, the crew had the hotel clerks swap themselves out with another clerk waiting below the counter. After a kind greeting to the hotel guest, the clerk went down under the counter to get something and then another clerk would come up and finish the task. None of the guests saw the new clerk because they were so focused on signing in. National Geographic conducted many similar scenarios demonstrating how veiling occurs when we are "hyper-focused." Few seem to be immune to this phenomenon!

I suggest that these same findings can be applied to our ability to see the things of God as well. The simple truths of understanding that can't be seen from one perspective can be revealed when our perspective changes. The Bible often reminds us that, in Christ, all things become new. I would also make the case that sometimes this idea can encompass that which was once hidden and later becomes visible. Although we may begin

> The enemy is highly skilled at the craft of distraction: Beware of his "slight of *mind*" deceptions.

to see the intangible truths, a narrow focus can inadvertently keep us from seeing the bigger picture that is being unveiled right before our eyes! The enemy is highly skilled at the craft of distraction: Beware of his "slight of *mind*" deceptions.

> **The traps of the enemy are usually quite invisible and each one is set with a distinct motive in mind.**

The Art of Distraction

Almost as passionately as God pursues us, the enemy of God chases us as well. Scripture teaches us that Satan sets traps for us.

> *And they may come to their senses and escape from the snare of the devil, having been held captive by him to do his will.*
>
> II Timothy 2: 26

Not only is the enemy constantly setting snares for us, but the traps of the enemy are usually quite invisible and each one is set with a distinct motive in mind. The Bible teaches that the enemy who was struck down from Heaven, roams the earth seeking whom to devour. Satan knows his time on earth is short and our time on this earth is about making a choice. If Satan can distract us long enough, he believes he can drag another soul with him into the realm of eternal suffering. If we can't see the handiwork of God in creation, the enemy is succeeding in distracting us.

I believe the devil is laughing wholeheartedly at us because we are so easily distracted. Many can't get past the first commandment "thou shall have no other gods before Me." If these

> **A sad tragedy is that we can be distracted from pursuing our *own* story and the epic adventure that God has written for us!**

distractions take all of our time, and we are not pursuing God, we are losing the battle.

A sad tragedy is that we can be distracted from pursuing our *own* story and the epic adventure that God has written for us! We wind up settling for watching other people's stories, factual or fictional. Sometimes our total social life revolves around what we'll see on TV next week. Moreover, as a nation, we seem especially interested in the "reality" shows of others. These shows depict dilemmas that we couldn't possibly imagine. We all have used distractions to pass the time or recuperate from a hard day at work or a challenging day taking care of our families. Our own lives can become so exhausting or unbearable that a distraction gets our minds off of our troubles for bit . . . before you know it, three hours have floated by, and we've been robbed of sleep, time with our family, or (more importantly) time we could have spent in God's presence.

> One question, in particular, has perplexed mankind for centuries: "What is the meaning of life?"

In fact, many people lose hours a day immersed in entertainment! Hollywood keeps cranking out epic dramas, which are not only entertaining, but have tremendous emotional appeal as well. We are strapped into a seat on a roller coaster ride that can thrill us beyond belief. It's quite a rush to experience great adventures without having to suffer any consequences or risk a thing—all for the price of admission!

Even today's authors will quote favorite movie scenes to help get a point across. It's so easy to do, especially when we have Hollywood investing millions in just a single action scene that communicates ideas so vividly. Why is so much energy being put into capturing our attention? Money of course! However, many writers, actors, producers and directors believe movies can also offer meaning and purpose to the lives of their viewers.

On the flip side, some people feel that Hollywood is nothing but a pit of evil; the devil's playground. Yes, when the content is destructive or meaningless, the enemy can enjoy lowering our morals, graying out what is right and wrong, encouraging the masses to become comfortable with sinful behavior and so on. Again, what's our response? Do we just sit there and take it in, or do we take the time and communicate to our children what the devil is truly doing here.[2]

One question, in particular, has perplexed mankind for centuries: "What is the meaning of life?" One interesting solution was presented in the sci-fi movie, *The Hitch Hikers Guide to the Galaxy*. After much anticipation, the two characters (and a multitude of others) were finally given a nonsensical reply to this age-old question. The most intelligent entity in the universe (a computer character in the movie they had created) merely responded with the words "42!" That's it? The number 42! That was all that this giant computer came up with after processing for seven and a half million years? It wasn't quite the ending anyone was looking for. In other words, there is no meaning to life. This is true for many. They are looking for meaning, but are not looking in the right place. As a result, they can find none. And "42" is better than nothing!

> If we have a desire to understand the things of God, then we should use what God left behind for us!

God gave us the answer key to any question we could possibly ever contemplate. He painstakingly inspired man to compose a holy blueprint, a guide, a love letter that we can count on. When we don't read the Bible, all we have is man's limited and finite wisdom. We gather what we can in the eighty or so years we are given, but it just cannot be compared to the wisdom and omniscience of an eternal God. If we have a desire to understand the things of God, then we should use what God left behind for us! The enemy has a dark motive and that is to keep us lost, deceived and in the dark. The non-sense of the number 42 is just that—a ride to nowhere at midnight.

God is not one to show partiality, but in every nation the man who fears Him and does what is right is welcome to Him.

Acts 10: 34–35

PC²

In a *60 Minute* segment about the life of Steve Jobs, co-founder of Apple, the question about his belief in God was raised. The biographer, Walter Isaacson, responded saying that Steve reflected his belief in the design of his devices. The words "on" and "off," as well as any on-off switches are not found on any of his products. The biographer stated that Steve did this because he was hoping that there is an afterlife. Either Steve feared the nothingness of "off" or he didn't know his destiny well enough to offer a stance or a position. Maybe he felt that this information was simply none of our business. Even so, Steve will be remembered for his genius. After all, that's why we desired to know his perspective on life's biggest question. I know that if hell (a place of incredible torment) was my destiny after death, I would want to pursue my beliefs to a place of safety. I would certainly give the subject of afterlife some very serious thought. Did Steve hold a belief that when life is turned "off," or when one dies, that's it—lights out? Could someone as bright and intellectual as Steve Jobs truly believe there is nothing more? Perhaps. Anyone can be deceived, especially those that seem the brightest. Along with great intellect can often come great pride. Of course for the believer, our answer is a resounding "NO!" Beyond a shadow of a doubt, there is so much more. Through God, our story has meaning, purpose and most of all, we have *everlasting* life!

If we don't make up our minds regarding what we believe, then the world may decide it for us. It's so

> **If we don't make up our minds regarding what we believe, then the world may decide it for us.**

easy to just follow the crowd, to bend toward what the trend of the day is. God is extremely clear that those who choose His way, the road less traveled, are a much smaller group than those who don't.

> *For the gate is small and the way is narrow that leads to life, and there are few who find it.*
>
> Matthew 7: 14

You will make known to me the path of life; In Your presence is fullness of joy; In Your right hand there are pleasures forever.

Psalms 16: 11

> **God is not partial when it comes to eternal issues. The decision is ours to own.**

No matter what status level we have or haven't achieved, God is not partial when it comes to eternal issues. The decision is ours to own. Be assured that if Steve wanted to make a statement of faith, people would listen as well as follow. Two questions remain: Did Steve not have the confidence because he knew he did not have the answer? Or, did the enemy cunningly and quietly use Steve Jobs' lack of a confessed public position on faith to plant seeds of apathy. Distraction is an important tool of the devil. Satan is an expert at hiding the truth. The enemy has over 5,000 years experience diverting man's attention away from God. He is the great thwarter and he will use every single resource he has to keep us from making a heart connection with God.

In battle, there is something called a "war table." It's designed to help commanders see everything more objectively:

> **The enemy has over 5,000 years experience diverting man's attention away from God.**

the enemy's position, the lay of the land, the obstacles, the plan, the specific maneuvers and the intended goals of those involved. Part of the agenda may include knocking out communications. The first side to achieve this immediately increases their chances in getting the upper hand, if not a sure victory. Communication is vital and the enemy knows it. The same is true on the spiritual war front. If we can't hear from God (or those who have a heart to bless others), then again we are at risk of losing the most significant battle in the history of the world—one that has eternal consequences.

> In our society, being "politically correct" can be interpreted in dangerous ways.

A popular (albeit slightly dated) mantra in corporate America encourages us to be "politically correct." This phrase is seen by some as possessing the ability to have a stronghold in "muzzling the messenger." In our society, being "politically correct" can be interpreted in dangerous ways. While we don't want to hurt anyone's feelings or be offensive in any way, when a message of hope can bring peace and life to someone, we should be able to provide that message to our fellow citizens. If the enemy is behind the use of this phrase, it has tremendous power. Hence, the name of this chapter section, PC^2. (The "^{2}" is used as a math multiplier. The initials "PC" means both Personal Computer and Politically Correct. Together, the story goes deeper than face value into areas of spiritual warfare.)

If indeed the enemy has a "war table," is there a tiny figurine with your name on it? . . . If so, what tactical distractive, deceptive or destructive maneuver is he employing?

> *Some people are like seed along the path, where the word is sown. As soon as they hear it, Satan comes and takes away the word that was sown in them.*
>
> *Mark 4: 15*

What Do You See Now?

Verse two of I Corinthians talks about the hidden and deeper things of God; those things which extend beyond natural or non-spiritual man's scrutiny. It's a spiritual matter to understand these deeper things that can only be given through the interpreter, the Holy Spirit. In verse ten, Paul points out that the Holy Spirit knows the heart of God and knows our heart too:

> **Paul points out that the Holy Spirit knows the heart of God and knows our heart too.**

> Yet to us God has unveiled and revealed them by and through His Spirit, for the Holy Spirit searches diligently, exploring and examining everything, even the bottomless things of God.
>
> I Corinthians 2: 10

In one of Big Idea's 3-2-1 Penguins! episodes called "The Carnival of Complaining", the host, Uncle Blobb asks a simple, yet profound question: "What do you see now?" Uncle Blobb reveals the positive and negative things that kids (and adults) often focus on. He demonstrates this while everyone is riding on a park monorail. The unpleasant host encourages them to see only the negative. Uncle Blobb spins everything negative. He would say that the ride was "too old" or "too boring" or that it was "just for very little kids." As expected, this started a complaining spiral that would make any parent dizzy! So the fun rides just kept passing by the group and the children did not take the opportunity to enjoy themselves. In a similar way, opportunities in our own life can keep passing us by. If we let long lines or the lack of cotton candy get us upset in the "carnival of life," we can be left with a drained countenance. God teaches us to manage our thoughts.

> **So God teaches us to manage our thoughts.**

He does this because He knows that we can!

> *It is not what enters into the mouth that defiles the man,*
> *but what proceeds out of the mouth, this defiles the man.*
>
> Matthew 15: 11

Opportunities in our own life can keep passing us by as we focus on the negative.

At times I find myself behaving similar to these children. But, through the Holy Spirit, I can recognize (or see) the sinful behavior that I need to reject. Indeed the "slight of mind" tricks of the enemy go far beyond our desires for cotton candy.

What do you "see" now?

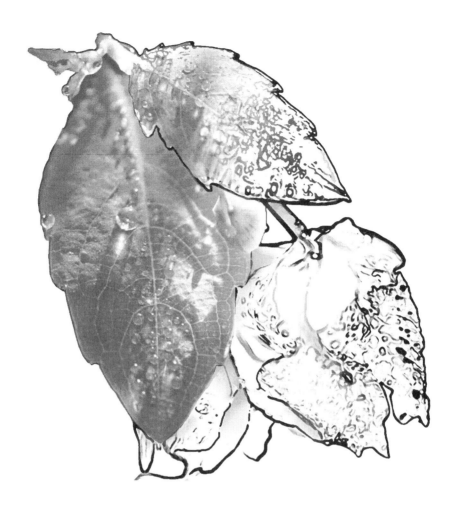

10
Obstacles That Keep Us from God's Passion

Unverified Information

This century has been dubbed the "information age." This brings blessings and responsibilities. The Internet brings our lives into hyper-speed: we now have virtual interactions with one another and immediate access to an overwhelming plethora of information. This dynamic allows technology and innovation of every sort to enter an exponential state of growth. Within the realms of invention and discovery, this can be very good. However, it can also create a portal for errors and misinformation to spill into our laps. Unfortunately, it is impossible to fully process the huge volume of information coming to us, literally, at the speed of light.

One of the necessary checks and balances we must incorporate into our filtering process is common, intelligent and intentional scrutiny. Information in and of itself has power and we are wise to be careful with that power. Technology can make, break or even destroy individuals, marriages and corporations. Even the government is not immune to the havoc that can result from the right information in the wrong hands. It can positively or negatively affect what will happen to people with their jobs, reputation, property and a host of other areas. Misinformation can misdirect and confuse while good information can have unlimited positive influence in our lives.

> **Technology can make, break or even destroy individuals, marriages and corporations.**

> **Misinformation can misdirect and confuse.**

It is of the utmost importance that we possess true integrity in our day-to-day living. If we are presenting information, we are to make every effort to be as error-free as possible. As human beings, we all have some element of bias. However, we must force ourselves to have an objective eye. We need to stop the information process "rocket ride" and take time to evaluate. When we do, we begin the process of setting aside unverified data or unproven hypotheses. Once this is done, it will help us to see creation in its purest sense, as God intended creation to be.

Be still, and know that I am God . . .

Psalms 46: 1

Leaving God Out

Science is one area easily affected by new technological advances, discoveries and theories. Some of these assumptions have been helpful in developing new theories, while others have been spun into old distorted scientific paradigms (a set of philosophical or theoretical frameworks). As years go by without acknowledging new discoveries that don't fit the paradigm, the scientific community and the world lose. They lose, not only important information that could shed light on current and past research, but even more importantly, God is ignored, belittled and not given the glory due Him. The latter is the most dangerous.

Some samples of the earth origin alternatives to Biblical Creationism may include these ideologies:

- "... mystical tinkering mechanism that miraculously spits out new..."[1]
- "... intelligence [other than God, such as aliens] must be involved for the extraordinarily diverse elements to work together.[2]
- The passing of time created...
- An epic explosion created...
- Need in and of itself creates...

> Today's findings can be in conflict with yesterday's assumptions.

Our "quest" for discovery should not ignore the hand of deity. If we are ignoring God's hand in creation, it would be good to search our hearts and ask ourselves why? Is there really a good enough reason? If there is a fear in recognizing God, then those fears should be addressed one at a time. Some of these fears may include:

- Recognizing God will be a hindrance or crutch in my discovery.
- If I recognize God, I will have to be accountable on what's right and wrong.
- If I recognize God, my peers will look down on me.
- If I recognize God, I will get lazy about going the extra mile.

- If I recognize God, it may jeopardize my job.
- If I recognize God, I may have to change my belief system.
- If I recognize God, I may have to eat a slice of humble pie.

When the essence of science is *true discovery*, our fears shouldn't be a factor. Like many fields of study, the discipline to go further—to go deeper in understanding shouldn't be at all mired! This is where I feel the popular "good intentions" excuse to justify leaving God as the Creator out goes awry and permits the rein of our unmanaged fears. The Bible teaches that we are *made* to glorify God, *not* to take the glory away.

> When the essence of science is *true discovery*, our fears shouldn't be a factor.

This information intersection (between science and religion) regarding the origin of life needs to *include* information, rather than exclude it. That means *all* information that has been validated by history and geology as well. And the Word of God has much validation (even though God doesn't need human validation).

Paul Garner brings interesting facts to life regarding this topic in the introduction of his book, *The New Creationism*. He states, "Major disciplines of science were founded by men of Christian convictions, such as Boyle, Ray, Hooke, Newton, and Faraday. These giant's were motivated by their spiritual beliefs! Like the astronomer Kepler, they perceived that in their scientific insights they were 'thinking God's thoughts after him'. Today, there is an embarrassed silence and a collective amnesia about the religious motivations of these men. This disconnect from biblical roots is atheistic in practice."[3]

> God doesn't need human validation.

Garner's sentiments clearly establish a tremendous obstacle which prevents many of God's children from seeing the intense love and passion that the great Designer has in His heart for each of us.

Fields of study which touch upon life's origin should at least include the *plausibility* that "God" created. Would the infamous Solyndra event (500 million dollars invested in a solar panel industry that failed) ever have happened if good science and/or thorough market research were initially presented? Were our leaders who invested this money on the public's behalf initially misled? Or, were regulatory or policy uncertainties the cause? Not to pick on just Soyndra, but GE put a halt on their energy project in this area as well. The point here is **not** to enter a blame game, but to make every attempt to include complete truth in all that we do. God is omniscient and is watching! His ways provide a benchmark that is reliable. If we drift from this, then whose benchmark do we use? Drifting has a price. It can be costly and keep us from true discovery. Garner was not alone in his ideas. Even the laws of our nation are biblical in nature. Three examples that are biblical: Thou shall not murder, steal or bear false witness (slander). We don't want to continue losing sight of evidence that doesn't fit into the current "PC" model. Observable evidence must always take precedence over theoretical assumptions.

Today's research and findings can affect the populous in ways beyond our human understandings! Acknowledging our Creator is the only choice. It opens the opportunity to broaden our understanding and see things on a much deeper level.

This God-Created concept helps us to avoid making those assumptions that are commonly made in science. As these leaps are proven wrong by today's observable evidence, it leaves society in a quagmire of misinformation muck. Once again we are prevented from seeing and experiencing God's passion for us.

As far as taking God out of education, Billy Graham, American Clergyman made a relevant statement. "America's founding fathers did not intend to take religion out of education. Many of the nation's greatest universities were founded by evangelists and religious leaders; but many of these have lost the founders concept and become secular institutions. Because of this attitude, secular education is stumbling and floundering."

Evidence of this is found in the surge of private school openings. This is something we should not stick our heads in the sand about.

> The God-Created concept helps us to avoid making those assumptions that are commonly made in science. As these leaps are proven wrong by today's observable evidence, it leaves society in a quagmire of misinformation muck.

Societal Misconceptions

To help simplify things, the terms "good science" and "bad science" will be used in this section of the book. Good science consists of witnessed, observable evidence without assumptions. Bad science is just the opposite—no witnesses, no observable evidence and it is loaded with assumptions.

Assumptions can be derived from:

- Theories that are based on generalizations, with a deductive (limited knowledge base) structure.
- Elegant consistencies within a synthetic (man-made) universe. Models are not reality, no matter how elegant.
- The assertion of the consequent: A model or set of equations has a finite (limited) domain of validity.
- One-sided, peer reviewed "journal published" content. Ossification (to become set in a rigidly conventional pattern,

a paradigm which is a philosophical or theoretical framework of any kind) of current assumptions.[4]

Scientific submissions regarding assumptions about creation would have a tough time holding up in today's court of law, since they are merely assumptions based on assumptions. They would, however, be accepted and published as **fact** should a single peer within a peer review, review it.[5]

> Complex genetic information in the form of genes and regulatory DNA *cannot* randomly evolve.
>
> Jeffrey Tomkins, Ph.D., Institute for Creation Research

As scientists continue to learn more and more about the genetic makeup of living things, it becomes increasingly difficult for them to fit their discoveries into the evolutionary model. In fact, these new discoveries actually negate evolutionary theory. A recent published discovery in the prestigious British journal Nature, proves such a point by stating that complex genetic information in the form of genes and regulatory DNA *cannot* randomly evolve.[6] Truth can be found in DNA.

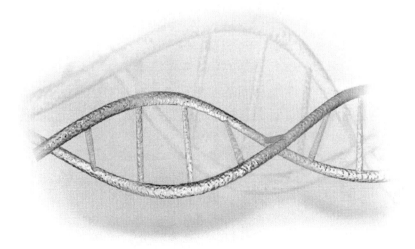

Below are a few of the "proven" theories that have been outed by more recent observable evidence. To read more of about these and other refuted theories, view "The Riddle of Origins Series", a DVD series by Mike Riddle and produced by *Answers In Genesis*.[7]

NOTE / Warning: This is a highly controversial area. I am not smart enough to solve it. Obviously, there are no eyewitnesses that can truly verify young or old earth origins. My hope is that issues of keeping a healthy *eternal perspective* will not be overshadowed by this or any other area of the creation mystery. God uses the mysteries of life for His greater and transcendent purposes.

- Were millions of years needed for fossil and coal development? Observable evidence found at Mt. St Helen's refutes this by demonstrating that it only takes two weeks for fossils and 20 years for coal.
- How about the millions of years needed for canyon development? Observable evidence found at Mt. St Helen's refutes this as well, it only takes five days.
- What about the animal species identified as extinct millions of years ago? Observable evidence of living "fossils" found without any evolutionary transformation. Some examples are the Coelacanth fish and Selakant.
- Did the human race evolve from primordial ooze? The failed Miller experiment produced the wrong enzyme needed for spontaneous life.
- Is the earth 4.54 billion years old? Modern observable data from the moon encourages views of a much younger earth.
- Did we evolve from monkeys? Archeological audits of Neanderthal man, Piltdown man, Lucy, etc. revealed that the skeletal findings were either falsified or grossly misrepresented on dig locations. Bone segment finds from different creations were miles away from each other.

- Is there really global warming? The un*supported* evidence and fact cover-up revealed by *Forbes* magazine states otherwise.[8]

Observable evidence shows earth temperatures cooling over the last ten years, and following a natural pattern of climate cycles.

"Einstein's relativity (good science), teaches that time, space, and light were not what we thought them to be. After relativity emerged, all of physics—even all of reality—was open for redefinition."[9] In the same way, new observable evidence should make evolutionary theories open for redefinition today.

The observable evidence easily moves Biblical Creation claims to a plausible status.

> *We demolish arguments and every pretension that sets itself against the knowledge of God, and we take captive every thought to make it obedient to Christ.*
>
> II Cor. 10:5

Denying the existence of a creator (which is the foundation of Genesis teaching), leads one to begin to question if the whole Bible can be trusted. If the first 13 verses of Genesis are false, then what does that say about the rest of the Bible? Once again, when we deny the Creator, we also turn our backs on His incredible boundless love for us, and reject the remarkable journey He has in store for us in this life—and the next!

> New observable evidence should make evolutionary theories open for redefinition today.

Ken Ham, founder of Answers In Genesis (AiG-USA), states: "Compromising Genesis has contributed toward the loss of biblical authority in our nation and helped open the door to the secularization of the culture."[10]

The Bible states that it is the Word of God, written by holy men of God.

According to John, "In the beginning was the Word and Word was God and is God." Genesis says that it took God six days to create everything To say it didn't, is synonymous with saying the Bible is nothing but a collection of myths—a book of lies that completely destroys any sense of biblical authority. (Some say that a day is like a thousand years, so time has a different meaning. This is another area of controversy. Again, we'll try to keep the topic as a God mystery for now). I believe in the Genesis account of a literal "day" translation. Believing in a long creation time subtly opens a "door of doubt" which has shown itself to be casting a shadow on cultural behavior . . . a departure from having a sensitivity of what is right and wrong.

> **If the first 13 verses of Genesis are false, then what does that say about the rest of the Bible?**

Answers In Genesis (AiG-USA), Institute for Creation Research (CRI), Creation Ministries International (CMI), Reasons To Believe, Biblical Creation Society, Creation Biology Study Group, Geoscience Research Institute, Creation Research Society, Earth History Research Center, WorldView Ministries, Creation Research UK, Creation Resources Trust, Creation Science Movement, Truth in Science and Creation Today offer hundreds of titles on biblical creation. And the interest on the subject is rising! Many are concerned about leaving God out of the science picture. Hence, one of the reasons for the rise in home school popularity.

> **Many are concerned about leaving God out of the science picture. Hence, one of the reasons for home school popularity.**

People are intellectually engaged. Our civilization is making extraordinary advances on all fronts by utilizing the latest equipment and technology. It would stand to reason that we would be open to new and astounding conclusions in that discovery process. As new molecular evidence is discovered that does not fit into the evolutionary model, it creates the need for a different model. I understand that in some scientific circles, there is a scramble for this new model. However, just stating it as the "Genesis account," would fit many of the missing puzzle pieces together. I also believe that the Lord would be pleased should we do this.

> It would stand to reason that we would be open to new and astounding conclusions in that discovery process.

Acknowledging God's role in creation is safe. He tells us too! God instructs us that His message needs to keep moving on. Our society attempts to shut it down by attempting to please all world religions. It just can't be done! God advanced humanity like humanity has never seen before in the short 200-plus years of America's existence. This is because the country was founded on Judeo-Christian principals. It never failed. We drifted away from it through our attempts of being "people pleasers." We are to make every effort to be "God pleasers."

> The "Genesis account," would fit many of the missing puzzle pieces together. I also believe that the Lord would be pleased should we do this.

Why did past scientists drop the "flat earth" theory for a new one? Observable evidence. Therefore, why do we continue to try and fit new biological and technological advances into a model developed in 1859? We are well beyond this path. This model is based on a theory developed before our "human discovery" of DNA and our invention of the electron and atomic microscope. The discoveries that we are making today are not only known by our Creator, but He is *thee* Originator of these as

> God advanced humanity like humanity has never seen before in the short 200+ years of America's existence.

well! To acknowledge God as Creator takes us all the way back to the beginning.

If the scientific community does not adjust itself and begin to apply all new, proven or observable evidence to their scientific research models—even if it does not fit into their existing models—then we, as a society have a responsibility to ask why.

> **What encouraged science of past history to leave out "flat earth" findings in the textbooks of yesterday?**

The most likely answer to the above question is fear. There is a fear of ridicule, fear of not being taken seriously, or fear of losing their position or standing within the scientific community. We are not to worry about what others may think of us. The Bible is clear on this: "If you confess me before others, I will confess you before the Father." When I run into a scientist or science instructor that recognizes God as their creator, I am reminded of the career risk they are taking and I esteem them greatly for their bravery and objectivity.

In his book *Ignorance: How it Drives Science*, Dr. Stuart Firestein, doesn't mince words on this topic: "What makes a scientist is ignorance. This may sound ridiculous, but for scientists, the facts are just a starting place." Dr. Firestein's point holds true as the sea of new information creates oceans of questions. He continues by saying that, ". . . thoroughly conscious ignorance . . . is a prelude to every real advance in knowledge."[11]

> **We all have quite of bit of yielding to do in regard to what we find as real truth.**

Scientists work very hard to be accepted into acclaimed schools. They pay for a high quality education and expect to receive high quality information. These individuals are using their gifts of knowledge to make scientific observations. In many cases, they didn't grow up with any sort of Bible teaching in the home, so there could have simply been a void of knowledge in Biblical matters. If a scientist would take the time to gain understanding, his or her world would expand beyond human constraints, and new advances in knowledge would be made.

We contribute to the secularization of our nation when we dismiss God as being relevant to science. The church has played a significant role by being irrelevant. The book *Already Gone* by Ken Ham and Britt Beemer (written with Todd Hillard) presents some revealing research statistics which deserve attention. The numbers are strong, but should be considered along with statistics of the past.[12] The book illustrates a strong symbolic sample of the short road to irrelevance. Charles Darwin popularized a philosophy that hit at the very foundation of the church (the Word of God). Surprisingly, Darwin was honored by the church and is even buried in the foundation of Westminster Abby.[13] *Already Gone*, although focused on a smaller window of research, offers several ways that the church could be more relevant in today's world. We all have quite of bit of yielding to do in regard to what we find as real truth. It is always good to bear in mind how God is working globally and, guess what? God's people are indeed winning![14]

> The fact that science has been very influential in many areas of life is undeniable.

I can't help but wonder, since the religious leaders imprisoned Galileo for his discoveries, but later learned that Galileo was right, that maybe the church was thinking . . . *We don't want to blow this one. Let's bury Darwin here as a public statement in our support of science.* It seems that Darwinism, evolution and millions of years, got "buried" in our children's brains as well.

Our approach to things of God can be very similar. For many, exploring the holy realms as reality can appear daunting. And, as in many fields, the knowledge required for understanding can be immense. Not that understanding things of a God is a "field" per say, it's more a journey of "personal faith." In any case, God defines levels of understanding by offering an analogy of milk (easy-to-grasp understanding) and of solid food (understanding that will take more study).[15] Whatever our life pursuits are, being "teachable" is a desired attribute.

Science is a great contributor to many good things in life. It's an area to not take for granted. The fact that science has been very influential in many areas of life is undeniable:

- Discoveries are helping with cures for diseases and improved health.
- Gene therapy restores meaningful vision to people with LCA and other forms of inherited blindness.
- Forensic sciences, investigation, DNA, engineering, psychology, drug chemistry. We have all benefited from scientific studies.
- Science and technologies can protect lives and livelihood from the effects of earthquakes, volcanic eruption, floods.
- Reliable water information for sustainable development are derived from science.
- Science has provided a basis for restoration of ecosystems throughout the country and world.
- Scientific innovation improves industry and technology.
- Energy and natural resources benefit from the exploratory advancements of science.

Yes, we are definitely grateful for the men and women who are making great strides in the field of science. However, the crucial issues of life should cause us to be prudent in how we evaluate the "scientific" interpretation of the world around us. Tough questions to ask ourselves are: If research is your field you may have to ask yourself which worldview am I utilizing to study the situation? How will this perspective influence my decisions? Which "applecart" will this idea upset? What happens if I lose my funding? Are there any alternate funding sources? These are not easy questions for anyone. Although the cost is high when it comes to making claims in the scientific world, it's also high if we sacrifice our personal faith. We all have a responsibility to be careful with issues that can affect outcomes, not only in this world, but also in the one to follow.

Standing up for what's right and true can be costly. I know because it made our life very difficult. We lost well into the six digits for standing up for our beliefs. It put some areas of our life into a tormenting downward spiral. However, I can now say that there were life lessons that I needed to learn and this was a sure way to get my attention.

Another societal misconception is that **all** scientists hold to evolution as fact. The *Select List of Science Academics, Scientists, and Scholars Who are Skeptical of Darwinism* was compiled by Jerry Bergman PhD. This group is considerably large, consisting of some 3,000 Darwin skeptics and is quite extensive. This study can be obtained at http://www.rae.org/darwinskeptics.pdf.

There is an additional list of 1,000 names which are unpublished. Dr. Bergman is honoring their request to not make their name public. Another list, from Physicians and Surgeons for Scientific Integrity (PSSI), lists 800 plus names of Physicians and Surgeons who dissent from Darwinism.[16] Because verification of the information on lists such as these can be difficult and time consuming, they are not created often. However, the information that is available is interesting, to say the least.

Weaknesses of the Peer Review Process

There are many weaknesses in the information exchange system that makes it possible to generate misleading information. One of which is *peer-reviewed* information. The information that science and technology banks on has gone through the professional peer review process and with it carries an assumption of "quality" or dependability. The *peer-reviewed article* is not only also printed or posted by a credible source, but is cited as well. As tidy as this looks, there are significant holes in this system that can cause a number of problems—all of which get the reader off on the wrong track.

So in short, if the information presented is misleading or distorted, it can go undetected. If the reader gets off track, good observable evidence can be ignored. If we are basing our work on a shaky foundation, it can easily crumble later on. Truth is always the best route.

We want good, solid information in our textbooks, so our kids don't have to wade through a sea of unverifiable and misleading information. We don't want to wait for the contributors to literally die off before we change the textbooks. This is the current path our society is on. And most importantly, it is a disservice to our children by hindering them to good discovery and learning all that God has for them to rediscover about His world.

I go into more detail about the peer review process as an obstacle and hindrance to good learning in an addendum in the back of this book.

> *Finally, brothers, whatever is true, whatever is noble, whatever is right, whatever is pure, whatever is lovely, whatever is admirable—if anything is excellent or praiseworthy—think about such things.*
>
> Philippians 4: 8

About Our Thinking . . . and Barriers to Learning

Communication is not complete if the intended message never reaches its destination. Whether its verified or unverified information, for God or against God, it's useless if it's not received and processed. William D. Winn, Director of the Learning Center, Human Interface Technology Laboratory at the University of Washington Educators states, "Children raised with the computer—think differently from the rest of us. They develop hypertext minds. They leap around. It's as though their cognitive structures were parallel, not sequential." Peter Moore expounds on this point in an Inferential Focus briefing by saying, "Linear thought processes that dominate educational systems now can actually retard learning for brains developed through game and Web-surfing processes on the computer."[17]

Today we are taking in information differently. As of 2012, YouTube stats include:

> **Communication is not complete if the intended message never reaches its destination.**

- Over 800 million unique users visit each month on YouTube.
- Over 4 billion hours of video are watched each month on YouTube.
- 500 years of YouTube video are watched everyday on Facebook and over 700 YouTube videos are shared on Twitter each minute.[18]

> **500 years of YouTube video are watched everyday on Facebook and over 700 YouTube videos are shared on Twitter each minute.**
>
> YouTube "Statistics" Accessed September 19, 2012

Our entertainment hours have also increased watching Hollywood's best. In each scenario, our physical activity involves sitting and staring. It's easy and it's fun! Today, we are shifting more toward using entertainment-based communications.

This section opened with a statement on what makes communication complete. However, what does it take to get information into the part of the brain for usability or practicality? Are electrical devices and entertainment the distraction of today? Nothing much has changed in 2,000 years. We still have the same two people groups. One group represents those that are life spectators and the other group, although much smaller in size, are life's participants. The spectators are just that: *watchers* of life. It seems as though when life calls for basic initiative engagement, there's a disconnect and the message isn't reaching the part of the brain for application. The critical thinking part of the brain can have an "unconscious-like" tendency.

> It would be safe to say that we haven't evolved much in 2,000 years!

This can encourage a "do nothing" response. Basic initiatives to engage safety or offer help just don't enter the picture! What were the distractions 2,000 years ago? Certainly not a drop down movie player for the second hump of the camel! They surely were not obsessed with games like *Temple Run*, or *Angry Birds*! Whatever the case then and now, it was a condition of the heart. It would be safe to say that we haven't evolved much in 2,000 years!

How does this create obstacles in experiencing God's passion? For me, it's like I am subconsciously embracing apathy. Why care about anything? After all, my brain is being entertained, right? According to my study of Ken Ham and Britt Beemer's book *Already Gone*, I am not alone. On any given evening, I think about all the faces which are glowing softly in the bluish light of the LCD screens and I am sad. People stare at their screens for hours, often missing opportunities to love each other, to reach out to others, or just to meditate on what God has done. God asks us to "Be still and know that I am God." Not "still" as a couch potato, but resting in His control of life's details and having an open heart to knowledge of our Creator.

> Not "still" as a couch potato, but resting in His control of life's details and having an open heart to knowledge of our Creator.

Experiencing God's passion means soul engagement, deep brain thinking and heart issue pondering. God says,

"Draw near to Me and I will draw near to you." God warns that He will reject us if we have not built a relationship with Him. God is talking to busy believers with mixed up priorities. God defines two types of people: those that know Him, and those that don't. If our life priorities are off track, then it's likely that the enemy is elated in the fact that we can't even get a handle on the first commandment. We have left very little room for seeing or experiencing God's passion.

The terms, "digital natives" and "digital immigrants" are used by our educators. You can probably figure which one describes you. Our teachers today stem from both groups. Both the natives and the immigrants are under pressure to entertain, or to liven up their subject matter. There seems to be constant pressure to make the information easier to grasp for the young "digital natives" in a desperate attempt to get ideas to penetrate growing minds.

> There seems to be constant pressure to make the information easier to grasp for the young "digital natives," in a desperate attempt to get ideas to penetrate their growing minds.

Dr. David A. Sousa, an international educational consultant and author of *How the Brain Learns*, states that some children who are currently labeled "learning disabled" may be more accurately described as "schooling disabled." Sousa adds that, "Sometimes, these students are struggling to learn in an environment that is inadvertently designed to frustrate their efforts. Just changing our instructional approach may be enough to move these students to the ranks of successful learners."

> Incorporating visual creativity in bible presentations is a great way to go!

Teachers of biblical concepts would do well to do the same. It could be fun! Incorporating visual creativity into Bible presentations is a great way to engage these "digital natives". By including the creative and biblical arts in your teaching, you can really help communicate the best message on the planet!

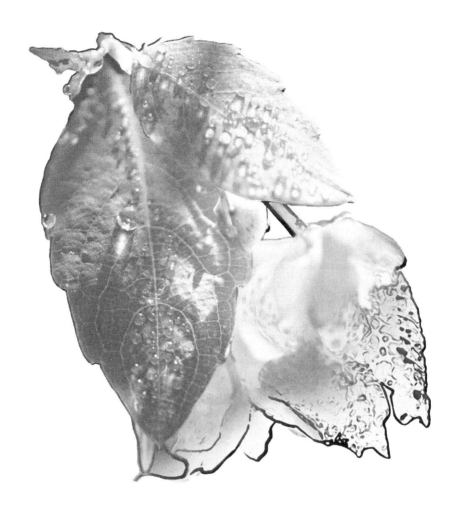

11
Transformers

People tend to resist change. Good or bad, change can be difficult. Good changes such as starting your dream job or the arrival of that baby you've wanted for so long, can be both exhilarating and terrifying. On the other hand, change that involves loss can be doubly difficult, even devastating. Death, separation from a loved one, changes in health, loss of a job, home, or any disruption in the status quo can have tremendous impact on our faith and outlook. Change affects all of us all the time. It's not a matter of *if*; it's a matter of *when* change will affect us.[1] Facing it head-on can be difficult. In fact, many of us refuse to do so! We easily find ways to practice avoidance. It's only natural!

In the movie *Transformers*, the producers, directors and designers go over the top in creating transformations. The 3D software technology, creativity and immense amount of patience combined, make these highly detailed transitions possible. The cars, trucks and newly invented beasts, literally turn themselves inside out several times (in explicit and exaggerated detail) as they reconfigure into a new shape. You see the inner stainless steel mechanics, gears, hydraulics, cylinders, cables, bearings, connectors and pulleys—and that's just what is recognizable! The movie is based on Hasbro's Transformer toy.

> **Change affects all of us all the time. It's not a matter of *if*; it's a matter of *when* change will affect us.**

However, theses transformers are no toys. The figures convert to Godzilla-sized, armed and armored, jet-packed robots. Your mind cannot grasp the amount of detail that is unfolding right before your eyes! The computer-generated graphics are very realistic. It's hard not to buy into this virtual reality *as* reality! All the radically moving parts end up tucking neatly together in each final transition. According to *Animation World Network*, it took 38 hours for ILM (Industrial Light and Magic) to render just one frame of movement in this movie.

You know that what you are seeing on the silver screen isn't real, but you buy into it anyway! The changes we witness on screen do not carry any tragic impact for us, unlike real life. It's great escapism! Seldom do we have a way to escape from the changes life brings. Our instinctual fear of the unknown can paralyze us from experiencing the great adventures that God has in store for us.

God created the "concept" of change before the beginning. God scripted changes throughout the silver screen of life. He knows our insecure state. God left us His Word and His universe. Since we are creatures of habit, God gave us some stable markers throughout His creation. He gave us the sun that continues to burn brightly and the seasons that come and go every year. He gave us changes that we can depend on! They comfort us by letting us know that there is order in the universe. The world keeps turning and God is on His throne. Yet, He is still a God of unexpected change.[2] These powerful reminders are the "transformers" of our reality that help us to keep our attention centered on Him. God even uses His creatures to teach us this. For example, the butterfly is the epitome of a transformed life! It turns from the lowly, almost awkward caterpillar to a beautiful winged wonder. The caterpillar magically turns into a chrysalis, its body becoming a puddle of goo, and emerges as something quite unrelated to its humble beginnings! These creatures seem to embrace change with unmatched beauty and grace. This is what the Father would like to do to us! He wants to transform us from our humble beginnings to the image of Christ! Ahhh, and what a joy it would be to Him if we could do it with the grace of the butterfly!

Change is a necessity for life-sustaining order. There are many useful things which He designed to spring forth from decay, such as coal, mulch and fertilizer. Even the processes of evaporation in the ever-changing hydrological cycle include changes in the water state, from vapor to liquid to ice.[3]

God says that we are to change, physically, mentally and spiritually. After our bodies expire, we change residence.[4] As creatures of curiosity, we gather as much knowledge as possible to help us through life. After all, knowledge is power! And through that knowledge, we grow in our relationship with God.

The opportunity to grow spiritually is the breath-taking element. It's what is truly vital in our adventure

> These powerful reminders are the "transformers" of our reality that help us to keep our attention centered on Him.

on this planet. If we would just throw ourselves completely into this venture, trusting completely in all that our Creator has in store for us without fearing the unknown, life (and beyond) would become so much more than we ever imagined it could be!

The Bible tells us that by the renewing of our minds, all things become new. This is a power-packed verse, to say the least. "All things become new" is a great promise. The Bible also tells us that God will "renew the years the locust has eaten." God promises to restore us from sinful, selfish creatures to images of Christ. With a "new mind," we get new perspectives. These new perspectives can give us hope in the future, even in pressing situations. Our new mind begins to reflect the mind of Christ. It gives us the insight and knowledge to change our circumstances. If we can have the mind of Christ, we can see things through His eyes, and begin to understand the lessons He wants to teach us and the path He wants us to take. We will begin to really grasp the power of prayer and how what seemed immovable, is now movable!

> Are we willing to risk it all as a demonstration of our desired friendship with God?

In another verse, God communicates to us that our spiritual growth begins with "milk." This milk is a metaphor for easy to understand principals and promises of God. We are not to stay on "milk" forever. Just as a baby is weaned from milk to solid food, so we too should mature to "solid food." How do we do this? By studying the Word of God for the deeper and richer things of God. This "milk-to-meat" change is how God designed our spiritual walk to be. We are to grow in the admonition of the Lord. If we are parents, we are to manage our children's metamorphosis by bringing them up in the way that they should go.

What about those decisions that we know we have to make that may result in painful changes? Sometimes making those changes seem impossible. They may result in serious repercussions if we make the wrong choice. We need to forge ahead, because God is bigger than all of them.

God is patient and not bound by time. He is in no hurry, but we are! We want the answers now! We want God to work in OUR time frame! We need to be patient and trust God to give us direction. We need to remember to surrender our thoughts and desires to Him. We are to pray for wisdom and God says that He will grant us it![5] He also tells us that there is wisdom in the counsel of many. One of the things that I like to do to help me get a clearer perspective on a decision is to write a list of pros and cons. I try and exhaust all of the possibilities of a decision that requires change. I include not only myself in the equation, but I also try and consider how my decision might affect others as well.

I take comfort in knowing my decisions are spiritually based. In addition, the verse stated by Jesus, "If you confess me before others, I will confess you before the Father" brings me solace. This, I believe is the ultimate crisis of life. Are we willing to risk it all as a demonstration of our desired friendship with God? If God is willing to give us the free gift of eternal life—complete with "heir status" as a child of the King, why wouldn't we?

Lee Strobel is a former atheist, lawyer and investigative reporter who realized he had to investigate the claims of Christianity, rather than blindly accept what he was taught. After almost two years of intensive investigation, he concluded that the evidence overwhelmingly supported the claims of Christianity. When Lee and his wife were married, she was an agnostic and he was an atheist. Later, when Lee's wife made the decision to follow Jesus Christ, it made Lee uncomfortable. He saw a change in her character that was both winsome and attractive. He went to church with other motives, thinking perhaps that she was in a cult. He heard the message of Jesus Christ articulated in an understandable way. From there, Lee utilized his background as an investigative reporter and went about investigating the claims of the Bible for nearly two years. The evidence that he gathered was so overwhelming that he realized *it would take more faith* for him to continue as an atheist than to become a believer in Jesus Christ![6]

As an agnostic college student, **Josh McDowell** believed that Christianity was worthless. But a group of Christians challenged him to examine the claims of Christianity on an intellectual basis. Instead of succeeding in discrediting the truth of Christianity, Josh discovered compelling historical evidence for the reliability of the Christian faith. As a result, Josh accepted Christ as his personal Savior and Lord, and he found his life changed through God's love and grace.[7]

Antony Flew was a contemporary British philosopher who was quite notorious throughout the world for his atheistic views (even referencing himself as such in one of his own book titles). Flew's arguments against God included: 1.The universe is eternal. 2. Life is a random process. 3. God is a self-contradiction. The academic world was set on its ear when, after much consideration, he converted to deism. He stated in an interview that the advancements of science itself reveal the integrated complexity of the physical world. Flew saw intelligence in the manifestation of life written in DNA, the transcription of DNA to RNA, RNA into proteins and the subsequent process of protein folding . . . He realized that intelligence *must* be involved for the extraordinarily diverse elements to all work together in harmony. Consciousness and reproduction consideration were influential in his subsequent conversion to theism as well. Rest in peace, Antony Flew.[8]

> God states that He disciplines those he loves.

These gentlemen all have exhibited faith by letting go of man's understanding and resting in their Creator's words.

> *Trust in the Lord with all your heart and lean not on your own understanding; in all your ways acknowledge Him and He will make your paths straight.*
>
> Proverbs 3: 5

> **When it comes to transformation, God is behind it!**

In God's mercy, God can orchestrate man's end. God says that He disciplines those he loves. Through trials, the disciple Paul learned the secret to the peace of God. He stated that he learned to be content in all situations. One interpretation of the Greek word content (or *arkeo*) means acknowledgment of God's control. In an email I received from Dr. Richard Swenson, his closing salutation read, "In His will is His peace." How true!

When it comes to transformation, whether it's the condition of a man's heart, life sustaining order or conducting a series of events so that just the right people show up at just the right time, God is behind it. It is a difficult task to try and understand God's orchestration of life events and heart transformation. Our finite thinking makes us incapable of really understanding.

If we could somehow capture the mystery, movements and thoughts behind God's actions, we would realize just how far short the detailed transformations in the movie *Transformers* really fall. It really is kids stuff compared to the workings of God. God doesn't orchestrate His transformations with virtual steel, gears and parts. He does it with His power and outstretched arm! Nothing is too difficult for Him!

> **If you hear His voice, do not harden your hearts.**

Part of our transformation to the likeness of Christ can involve divine tests.[9] Hebrews 4: 7 states, ". . . If you hear His voice, do not harden your hearts." Additionally, Hebrews 3:19 warns, "So we see that they were not able to enter, (rest) because of their unbelief."

> *Do not be conformed to this world, but be transformed by the renewal of your mind, that by testing you may discern what is the will of God, what is good and acceptable and perfect.*
>
> Romans 12: 2

God demonstrated His awesome power through many miracles, but the Israelites' faith was conditional. The Israelites wanted to live on their terms, not God's. God showed them the parting of the sea, the pillar of fire and He provided meat from Heaven and water from rocks! None of these incredible miracles softened their hearts toward God. I believe this is why God isn't so obvious with His miracles today. Heart decisions need to be thought through and, as Paul points out, learned. There needs to be a surrendering of the will and an acknowledgement of His control, no matter what the circumstances or conditions. God is God, no matter how we justify our thinking.

> I believe this is why God isn't so obvious today with His miracles. Heart decisions need to be thought through.

We are not to fear any of God's transitions. A practical way to get destructive fear out of our lives is to gain understanding and knowledge. Transition can be interpreted as a crisis time for many people. On the contrary, the Chinese definition of crisis is: "a time of opportunity." Under the loving care of our Heavenly Father, we are provided exactly that! We are given the opportunity to get to really know our Heavenly father who *is* within our reach.

Often life transitions come up like a storm, tossing us from here to there. The Bible uses storm imagery several times to reveal certain aspects of God. When the disciples were tossed about in the boat, they were filled with fear for their very lives. Talk about a crisis! But God empowered Jesus to calm the storm with his hand and just by uttering a word or two, the sea and the disciples were calm once again. When God spoke to Job in the midst of rain and thunder, Job was enduring a stormy sea of his own. Once the storm retreated, Job's life would be used to glorify his Father. Like Job,

> God is God, no matter how we justify our thinking.

when we are worried or anxious, we can call upon God and we will receive a message of restoration, love and passion. It is interesting that the book of Psalms, which offers a very soul-soothing message, follows the book of Job with all of his trials.

Job is a great example of how we don't necessarily embrace God's plan for our lives and the lessons He wants to teach us through trials. We live a linear existence and it is difficult for us to see the end of our suffering or to believe that it will result in anything good! We do not like unexpected changes and detours. We prefer a clearly outlined map that takes us around all the troubles! However, it is often the trials that push us to let go more and more of ourselves, and obtain more and more of Christ. These verses have helped me in keeping a "big picture" perspective.

> *Like Job, when we are worried or anxious, we can call upon God and we will receive a message of restoration, love and passion.*

He frustrates the plotting of the shrewd, so that their hands cannot attain success.

Job 5:12

Blessed is the man whom God corrects: so do not despise the discipline of the Almighty.

Job 5:17

For You have tried us, O God; You have refined us as silver is refined.

Psalms 66: 10

These trials will show that your faith is genuine. It is being tested as fire tests and purifies gold—though your faith is far more precious than mere gold. So when your faith remains strong through many trials, it will bring you much praise and glory and honor on the day when Jesus Christ is revealed to the whole world.

I Peter 1: 7

God knows we are uncomfortable with change. We shouldn't be, but that is the way we are. We should ask why this is the case.

From the time the two cells become one, a heavenly transformation was underway. Today's new scanning technology is able to focus in on the very beginning of human life like never before in detail, clarity and color. Alexander Tstiaris, Associate Professor and Chief of Scientific Visualization of Yale, produced an eye opening conception to birth video.[10] Tstiaris' work reveals epic transformation unlike anything anyone has seen before. Tstiaris reveals his reaction to what he found while working on the project:

> "The magic of the mechanisms inside each genetic structure saying exactly where that nerve cell should go—the complexity of these, the mathematical models of how these things are indeed done are beyond human comprehension. Even though I'm a mathematician, I look at this with a marvel of how did these instruction sets not make mistakes as they build what is us. It's a mystery, its magic, its divinity."

As an example, Alexander shares what he learned out about collagen, which is found in our skin and hair. "The only place that the collagen *changes* its molecular structure is in the eye. It becomes a grid formation which creates the transparency needed in the covering."

We are God's biggest and most complex transformer!

Transformation is in our DNA's original design! It just happens to slow down after age 16 for gals and age 18 for guys, and then stops around 25 (but the change process is still engaged). We are God's biggest and most complex transformer! Is it another design motive to push us to lean on Him?

The mysteries of change are certainly ponderous. We have looked at change in its many forms and we can see how God uses change to transform us and push us towards Him. God-wrought changes can put the pressure on each of us to get to know God better and reach new heights in understanding. We are pushed to reach for deeper knowledge and deeper faith. People from all walks of life are called to pursue God, especially in prayer. Whether you are the President of the United States, or a mother up late at night with a colicky baby, you may find yourself getting down on your knees and praying for strength to handle your current difficulties with humility, patience and grace. Through change, we become fully aware of our limitations and realize that we need help from on high! When our backs are against the wall, when we are out of strength and answers, we finally call on our Creator and pray. That's just where God wants us! He wants us to depend on Him through faith and fully trust in Him—every single day!

Our faith journey is filled with transitions . . . We move from a human center of understanding, based on man's point of view, to a new center of understanding based on God's point of view—and God's point of view can give us an eternal perspective! It takes some getting used to, but as we pursue knowledge of our Creator, His Word performs its work within us (I Thessalonians 2:13).

That's very exciting! When all things become new, we have access to peace in all situations, perspective in chaos, order in disorder and hope of promises to come.

> *All Scripture is inspired by God and profitable for teaching, for reproof, for correction, for training in righteousness; so that the man of God may be adequate, equipped for every good work.*
>
> II Timothy 3: 16–17

12
Hope in the Passion Creator

In his observations of conception, Alexander Tstiaris brings attention to a molecular communication. Alexander states, "when a fertilized egg burrows and connects itself to the side of uterine wall, cell to cell communications commence. 'I am here to stay; plant me.' The estrogen and progesterone cells hear this call and respond!" The cells have been quiet for many years in the women's life until this call. The cells were programmed and designed to respond upon the call from the fertilized egg.

The instruction sets are written in the cells! No wonder we don't want to read the instructions when we are trying to put something together! It's innate with us! The mother's womb is designed to offer this new life everything that's needed! Warmth, protection, circulation, nutrients, oxygen and a host of other life giving elements are in place to do their thing at just the right time! The Psalmist writes, "You knit me together in my mother's womb" and in Matthew he states that "He knows the number of hairs on our heads."

With the help of the Holy Spirit (which we can have by just asking through prayer), we can wrap our minds around Jesus Christ's presence at the beginning of the universe and our own inception. His loving care for us is intimately written in all of existence. *When Jesus creates, it's all about us. When we see Jesus' hand in creation, it's all about Him!*

> When Jesus creates, it's all about us. When we see Jesus' hand in creation, it's all about Him!

Like the fertilized egg securely embedded in the mother's womb, so too can your hope be embedded in what scripture calls the vine. "I am the vine, you are the branches; he who abides in Me and I in him, he bears much fruit, for apart from Me you can do nothing."[1]

Just like our life giving cells that know exactly what to do at just the right time, all that God has planned for us is set in motion when we "commit" to Him. John 3:36 reads, "He who believes in Me will have eternal life." The Greek definition of believe is to "commit and adhere to." John 1:1 states, "In the beginning was the Word, and the Word was with God, and the Word was God." Committing and adhering to Jesus, who is the Word of God, is the only safe place to rest your hope!

> All that God has planned for us is set in motion when we "commit" to Him.

The Other Side of Heaven

Because of sin, humanity is not perfect. We're sad when we see people who claim to be Christians, but act like hypocrites. When something terrible happens to a friend or family member, we wonder how a loving God can permit such bad things to happen to good people. Yes, we will see evil on earth; we arc on the wrong side of heaven. Until Jesus comes again, there will always be evil on the earth. It is hard for us to except, especially when we get a glimpse of our loving Father, and naturally expect His love to permeate all earth and humanity. But humanity is fallen and the earth cursed. However, through the sacrifice of Christ, we are not alone. The Father has given us a choice: accept Christ and find comfort or dismiss Him and walk alone. Until Jesus returns, evil will reign on the earth, so we must fix our eyes, our hope and our trust in Him.

> Who would think that our home would be a dangerous battlefield for our soul?

This side of heaven is our current residence. Who would think that our home would be a dangerous battlefield for our soul? If we believe in God, we have to understand that He is a just and transcendent God. When it comes to our souls, God plays hardball. His love for us is not about making us happy, but about making us perfect. Earlier we discovered that God wants us to have an intimate, trusting relationship with Him. Satan would like us to believe that God doesn't love us, that God has turned His back on us because God is not giving us what WE want. Satan wants us doubting God. Satan wants us to feel abandoned, alone and cold, sulking in the

> Satan wants us doubting God. Satan wants us to feel abandoned, alone and cold, sulking in the corner and muttering that life isn't fair.

corner and muttering that life isn't fair. Satan used this same strategy in the Garden of Eden and even tried this tactic with Jesus in the desert.

I was there not long ago. Satan was winning and I couldn't see it. I was on a downward slope, heading into a valley of despair and ultimately, facing a spiritual death. In fact, I was so low that it could have led me to my physical death. It wasn't until this moment that I recognized the deep trust and hope I needed to have in my loving God. I remember reminding myself that I could not make one simple creation, like a blade of grass, from absolutely nothing. It reminded me of the design of creation and that my sustained existence is a concern of my creator. The thoughts of my Creator kept my perspective and kept me alive.

> I could not make one simple creation, like a blade of grass, from absolutely nothing.

The trial really challenged me in issues of faith. I had years of volunteering for church and Para-church organizations under my belt. I had done award-winning work for many well known blue-chip companies. I believed that this work entitled me to God's blessings in everything. Little did I know that a death by 1,000 cuts was just around the corner. Several key business contacts had relocated. I kept loyal employees on the payroll too long. Design software miraculously turned personal computers into design studios and the economy came to a halt. I did not have the business acumen or training to turn this around quickly. Even if I had, I don't know how much it would have helped because so much of it was out of my control. What I also found difficult was the lack of understanding among non-business owners and business owners in other markets.

> Reading and memorizing verses like these helped to give me the right perspective.

When there is a downturn in the economy, design communication and advertising firms are usually one of the first industries to feel it. Needless to say, I felt very distant from God and the world! I took it all personally and began to feel like a modern day leper. I remember the local newspaper carrier forgiving our debt to us. At the time, it felt like the most merciful thing anyone had ever done for me. Life did not feel fair.

The behavior that helped me get through this was scripture reading and memory. Verses like: it rains on the just and unjust, He who began a good work in you will perfect it, He hears our cries and answers them, gave me hope. He gives good gifts to those who ask, and that I am more valuable than birds and God takes care of them. God is my refuge and my strength, a very present help in times of trouble, are verses that gave me reassurance. Reading and memorizing verses like these helped to give me the right perspective.

> God is the "living water" that quenches the thirst for those that recognize that they are thirsty.

The incredible acts of kindness by loving faithful believers, along with the prayers of many, were much needed reminders of God's care and provision during this time. These reminders helped to hold my faith in a surrendered state. Indeed, God's ways are not our ways. If going through trials is the price we must pay to be able to "see" the passion that God has for us, then the price is fitting! The promised peace that Jesus offers which transcends all understanding, will guard our hear hearts and minds in Christ Jesus.

But the tough times can be difficult! The desert's heat is beyond uncomfortable. The thirst is unbearable. This desolate place can carry unimaginable loneliness and despair. Adam and Eve were once in the lush garden of God's love, but they made their own choice and were turned out to work the barren land with sweat and tears. Because of that, we now live by the sweat of the brow until He comes again. The desert, though it can be a place of hardship on this side of Heaven, is also a place where God meets His people. God is the "living water" that quenches the thirst for those that recognize that they are thirsty.

His plan includes life coming at us with full force. Yielding to the Creator is the first step. Next, we need to except that life events can bring us closer to God. It's up to us to engage with Him. This is what makes the journey a challenge and redemption exciting! It's all worth it if we look to our *real* home—on the other side of Heaven!

> His plan includes life coming at us with full force.

A Passionately and Creatively Prepared Place

Prior to Jesus' ascension into Heaven, He told us that he would go and "prepare a place" for us. The operative word here is "prepare." The Preparer is the one who ushered in the universe. He has been at this construction project for almost 2,000 years. Heaven is described in specific terms: "No eye has seen, no ear has heard, no mind has conceived what God has prepared for those who love him."[2] OK, this is inconceivable. In the book of Revelation, the Apostle John speaks of the Revelation of the Lord Jesus Christ. He tells us of the New Jerusalem, a city with roads made of gold, walls made of rubies and gates made of pearls. Not having Satan around will be joy enough for me. But wait there's more! God promises that in His mansion there are many rooms. Scripture also gives physical dimensions of the new kingdom to come.

I believe we get a better idea of Heaven when we look at our Creator with His "beyond understanding" attributes. We know He operates outside this dimension and is separated from our concepts of time, space, mass, and energy. He

> Jesus is not bound in any way by what we see, hear, touch or sense. He is Master of the microbial and God of the galaxies.

used a handful of elements to create life. Scripture states, "He is before all things and in Him all things hold together." Jesus is not bound in any way by what we see, hear, touch or sense. He is Master of the microbial and God of the galaxies.

There are so many questions I have about our real home. Will God have a new color we haven't ever seen? Will He use a new raw element base? Will there be new musical instruments or will our hearing be newly tuned so that we will be able to hear sounds that have always been around us? Will we have any of His divine abilities, such as walking on water or turning water into wine?

> Will God have a new color we haven't ever seen? Will He use a new raw element base? Will we be able to hear sounds that have always been around us?

God told us there was going to be a banquet upon arrival. Well I am glad to hear that there will be food! What is a church gathering without food? I can't imagine the flavors. Talk about flavor enhancers! Will there be new types of veggies or fruit? And how about steaks? The most exciting thing my finite mind (and stomach) can hope for would be brand new flavors of ice cream! If you think Ben and Jerry's have some neat flavors, hang on to your halos!

I get such a kick out of smells, like the smell of fresh ground coffee beans in a coffee shop! Will there be a new heavenly bean to grind? My wife is a big fan of teas. Put on your creative cap and try to name a few of the terrific tasting teas that wait for us at the banquet of a lifetime!

> Our Master Creator is "beyond" what is imaginable!

There have to be flowers in Heaven! And I hope our nose smell receptors can hold onto a scent longer than five seconds! The scent of blooming carnations, roses, lilacs, and perhaps brand new hybrids, all in bloom in the mansion's beautiful green house . . . all year long. I can't imagine it!

Our Master Creator is "beyond" what is imaginable!

A Passionate Prayer, A Passionate Praise

My son is a member of the "digital native" group that was discussed earlier. My wife and I have been very comfortable with praying often. For us, it is a way of life. Our son is at an age where we feel he needs to start to go a little deeper in his faith walk. The challenge is before us. What can he relate to that shows the power of prayer?

Another hurdle to overcome is that my son's current belief is that God doesn't answer prayer. He witnessed what we went through, our struggles and frustrations. It is written,

> *I will destroy the wisdom of the wise; the intelligence of the intelligent I will frustrate.*
>
> I Corinthians 1: 19

After a few years of divine hammering, I was just beginning to learn about the "obey" part and that God expects me to actually practice what I am learning! Unfortunately, we could only shelter our son so much from the harsh realities of life. And, when my teenaged son asked for things, such as a fishing trip, a dirt bike and new expensive clothing and didn't get them, he blamed God. And when the family had to sell the four-wheeler, it was the end of all things from his teenage perspective.

We all have preconceived ideas about deity. These ideas stem from our childhood. Some of us had a lifetime gathering and surmising our experiences to reach our current belief system. Others may come to the

> **God expects me to actually practice what I am learning!**

> We all have preconceived ideas about deity. These ideas stem from our childhood.

table after experiencing destructive situations or may come from dysfunctional families. Sin can steer our ship into these rocky shores, but often, we just aren't vigilant enough to avoid the dangerous shipwrecks. Unaddressed hurts inflicted by others can become the bitterness that eats away at our hearts, like the saltwater waves that hammer the deck of a ship. Whether a victim or an offender, justifying and solidifying our stance can be like an anchor that weighs us down. We are no longer free. We are unable to sail. Satan is pleased that we cannot see the lighthouse of Christ that guides us home. We are prevented from being as God intended us to be: His pure, undefiled, God-breathed Creations.

God wants us to set a course for a place where we are clean and free from sin—humbly bowed before our Captain and life-Designer.

Some of us are literally physically hardwired to our negative experiences. Science can actually "see" the hardwiring that happens on a molecular level in our brains.[3]

> What's alarming about this is that when we were young children with untrained minds, we were inadvertently wiring ourselves into our belief systems of today.

These rewiring agents can be negative, repetitive self-talk stemming from bad life experiences. It also can be from physical or verbal abuse from peers, friends and caregivers. Science shows the rewiring of our brain cells physically form new connections of self-perceptions. Whether the interpretation is right or wrong, our brain cells shift, build and connect to accommodate our repetitive thinking patterns. We literally hardwire ourselves to "our understandings" of situations, experiences, good, bad and even the nonsensical.

What's alarming about this is that when we were young children with untrained minds, we were inadvertently "wiring ourselves" into our belief systems of today.

> *Train up a child in the way he should go,*
> *Even when he is old he will not depart from it.*
>
> Proverbs 22: 6

As adults, how many have had any training in high school (or even in college) on how to handle social media slander, sexual violations, abuse, neglect or addictions in any form? Our changing society is leaving children and adults exposed and unprepared for events and situations such as these. Deep heart-hurts are left unattended. God desires to address and heal these. One example of this hard-wired concept is explained by the admonition of the Hebrews in the Bible:

> *Look after each other so that none of you fails to receive the grace of God. Watch out that no poisonous root of bitterness grows up to trouble you, corrupting many.*
>
> Hebrews 12: 15

> *See to it that no one comes short of the grace of God; that no root of bitterness springing up causes trouble, and by it be defiled;*
>
> Hebrews 12: 15

I have used two different translations of the same verse to give a better understanding of what the author is telling us. Psychologists say that ninety percent of counseling cases involve unresolved issues due to unforgiveness. The word "root" is an appropriate metaphor for what bitterness does to our souls. A root relates to the subterranean or below the surface; what is not visible but spreads relentlessly. As I have mentioned before, we have a farm and have had our share of weeds. I have spoken of the Canada Thistle, but let me tell you about bamboo. It spreads through its root system. You pull one up and can follow the connecting root to the next plant and the next and so on. That is how bitterness spreads in our hearts. If we don't stop it in its tracks it will spring up and cause us all kind of trouble in many different areas of our lives. Have you ever met a bitter person? They cannot find joy in anything. They become mean and spiteful. The bitterness can destroy relationships and be the cause of a loss of a job. God gave us a very appropriate analogy in the use of the word "root."

> Should our fallen perspectives dictate our eternity and have power to influence our relationship with God?

Our Adamic nature (given to us upon Adam and Eve's rebellion) is bent toward sin, resentment, and bitterness. We seek out the negative and untrue. We are fascinated with reality shows that really do a good job of demonstrating our fallen nature. We give a sigh of relief and swipe our hand across our brow when the cop and the flashing lights pass us and pull over another car! We say to ourselves "Whew, glad that's not me!" This is a natural response for our fleshly nature. But our flesh is not to rule us! We are to respond in a way contrary to our nature. We are to respond like Christ. What would that look like? "I am thankful that that speeder did not hurt someone or get in an accident. Maybe the officer will exhibit mercy, and maybe the speeder will have learned a lesson." The fight begins because the Adamic nature has life-altering roots!

The question arises: Do these fallen perspectives dictate our eternity and have power to influence our relationship with God? The roots of negative experiences can keep us from seeing the truth and all that God has to offer us. Sound Christian counseling can restore a forgiving nature.

Jesus' passionate prayer when being put to death was "Forgive them Father, for they know not what they do." This prayer has protection in its intent and the act has passion in its purpose. Jesus knew of the persecutor's Adamic sin nature. He knew of the personal baggage and the condition of the hearts of all involved in His death. Jesus knew that they were lost children making wrong decisions .

Jesus saw His persecutors' frailty. He saw past their condemned condition, their rags of guilty works. Maybe they were not *redemption worthy*, but Christ saw them as *worth redeeming!* Christ entreated the heavenly Father's compassion and understanding and asked Him to forgive his killers for their horrific deed, as Christ had not yet died to pay the penalty for it. Is it possible for us to acknowledge that we are just as frail as our predecessors, that we are just as depraved in our hearts? To acknowledge this is to realize we need a savior.

> Maybe they were not *redemption worthy* but Christ saw them as *worth redeeming!*

God tells us that above all else, we need to guard our hearts. Our hearts were initially designed to love God. To pray passionately, we need to get real with ourselves and care about God's initial design of our hearts. His desire for us was to have hearts that are sensitive to the things of God and to Christ's soul cleansing work on the cross.

Through His stripes we are healed.

Isaiah 53: 4–5

The penalty for turning our hearts away from God and choosing sin over Him is death. However, the Bible clearly demonstrates that from death, new life can spring forth. There is a complete reversal of the horrible finality of death and it becomes a glorious renewal!

In the Jewish tradition of the Passover, the blood of a sacrificed lamb was painted on the doorways to protect them from the angel of death as he visited the land of the Pharaoh. Moses' people were saved.

In addition, the frequent practice of sacrificing a lamb at the temple was done to cover the sins of the people. The shedding of pure, holy and innocent blood satisfies the penalty of all sin: past, present and future. Likewise, Jesus was sent as mankind's perfect sacrifice. Jesus satisfies the debt that is described in John 3:16. By accepting this free gift, we experience healing in our relationship with the Father and we are restored once again.

> God knows of our hardwiring! He knows exactly how each brain receptor is connected, rerouted and/or disconnected.

We pray to know God. I John reads, "To know God is to fear Him and obey His commands." The Hebrew and Chaldean definition of fear (yare) is to have an awesome respect or reverence for something or someone.[4] Although obedience can be a tough endeavor, God knows of our hardwiring! He knows exactly how each brain receptor is connected, rerouted and/or disconnected. Paul writes that we should work out our salvation with fear (yare) and trembling. All our bad habits and attitudes are not necessarily changed in an instant. Sometimes God wants us to work things out with Him at our side. This is often how God teaches us His character and builds that character in us! What is changed in an instant is our eternal destination! We now have an eternal home in Heaven and we are privileged heirs and children of the King!

If we are talking about prayer, why is it necessary to talk about forgiveness? Sin is the divider between man and God. The prayer connection is severed until the sin issue is acknowledged. Understanding that sin separates us from God is the first step to reconnecting with Him.

> *But your iniquities have made a separation between you and your God, And your sins have hidden His face from you so that He does not hear.*
>
> Isaiah 59: 2

We want our prayers to be unencumbered by man's fallen state. God wants us to know that He didn't create us to be living in sin.

> *For all have sinned and fall short of the glory of God...*
>
> Romans 3: 23

We now have an eternal home in Heaven and we are privileged heirs and children of the King!

God originally created us to be free from sin—no evil desires, greed, disease, covetousness, etc. Adam and Eve, in our stead, made the choice that we all would have made eventually. They chose to disobey God and the penalty for their disobedience was to know good and evil. We know evil and know it all too well. This choice separated Adam and Eve and all of mankind from God. They no longer walked in the garden with their Father. The relationship was broken.

Even in the midst of this broken relationship, God still gave man a way to connect with Him—through prayer. This thread of friendship is found throughout the Bible. Through prayer we have a way to communicate to the Creator of the Universe, the One who put the sun and moon and stars in their place. Why not follow the communication instructions that He left for us?

> *The time is fulfilled, and the kingdom of God is at hand; Repent and believe in the gospel.*
>
> Mark 1: 15

God made you. God made me. The Bible teaches that He knows what we are going to ask before we ask. He is omniscience in its full capacity. God desires to communicate with us—the ones that He was thinking of before the beginning. God wrote your name in the Book of Life!

God wants to hear from you. The verses below can help you as you pray. See them as God's personal messages to you. Meditate on them. They are his love letter, written just for you.

> *For God so loved the world, that He gave His only begotten Son, that whoever believes in Him shall not perish, but have eternal life.*
>
> John 3: 16

> *Let your gentleness be evident to all. The Lord is near. Do not be anxious about anything, but in everything, by prayer and petition, with thanksgiving, present your requests to God. And the peace of God, which transcends all understanding, will guard your hearts and minds in Christ Jesus. Finally, brothers, whatever is true, whatever is noble, whatever is right, whatever is pure, whatever is lovely, whatever is admirable—if anything is excellent or praiseworthy—think about such things. Whatever you have learned or received or heard from me, or seen in me—put it into practice. And the God of peace will be with you.*
>
> Philippians 4: 5–9

You can be upset, hiding something or ashamed . . . it doesn't matter! He knows all about it! Prayer crosses into the spiritual dimension of life. God casts a shadow over our right hand, so He is nearby and listening!

Just as I was finishing the final chapter of this book, I had a very unusual dream. There was a slightly grayed haired gentleman who seemed to be caught up in his sin. He was handling and shuffling all sorts of very sharp objects. They were in a variety of shapes and sizes; some were shard-like, made of steel or glass, and many of them had handles on them. He started to throw some of these—impaling them into the walls and woodwork. In my dream, I felt threatened and so did the other eight to ten people in the room. The tension rose greatly.

The gentleman seemed to be a father figure at first—someone you could trust. I sensed an intense frustration as he threw the sharp objects. The incident could have easily caused death or injury, but for some reason it didn't cause either.

I was able to detect remorse residing deep in his eyes. Cautiously I asked the man, "Do you feel sorry?" I later interpreted this part of the dream as a deep struggle within, a conflict of the soul that has been buried for a long time, but has now been stirred. There is no apparent way out. The frustration manifested itself with the burst of rage. The gentleman looked at me intently, and his eyes told me that he was remorseful. I then simply said, "Jesus loves you." At that moment the man seemed to surrender his will and became peace filled. The anxiety left the room. Suddenly I awoke, and wondered what were the implications of this dream. I had the impression that these images would have significant meaning for someone else. To be honest, I believe the significance and interpretation of it lies in you, the reader. This is why I am including it in this book. How does this dream speak to you? My wife Tammy gives her interpretation:

"We all have the broken pieces of our lives. They lie on the floor like dangerous shards of glass that we must carefully step around. We wish we could get a handle on them. How much we would like to just be able to grab them and toss them aside. But they don't usually come with handles. However God can use these broken shards as tools for us to grow and learn by. He can put the handles on them so that we can pick them up and examine them and toss them aside for Christ. Often God uses them to draw us to Himself, to get us to finally 'give up' and surrender it all to Him."

One thing that struck me was that none of the people in the room were hurt. They were somehow shielded. The gentleman hadn't yet lost his perspective on what is really valuable on this side of Heaven... *life* itself!

Never will I leave you; never will I forsake you.

Hebrews 13: 5

If you have prayed the previous verses back to God and the prayers are from your heart, you have prayed a prayer of salvation! You have been saved! Welcome home sister! Welcome home brother! Philippians 4: 9 instructs us to put what we've learned, received or heard into practice! You can start by seeking out a Bible believing church to attend. Communicate to a church leader where you are at with spiritual matters. Share any stumbling blocks that you feel could be hindering your perspective. He or she should be able to point you in an appropriate direction. You may be surprised on how setbacks can be turned into strengths with God's real-world help.

The praise is simple but still just as real and heartfelt as *your passionate prayer*. Thank God for your free gift of salvation, the Kingdom to come and His promise of everlasting life. God is rejoicing over you!

> *The Lord your God is with you, he is mighty to save.*
> *He will take great delight in you, he will quiet you*
> *with his love, he will rejoice over you with singing.*
>
> Zephaniah 3: 17

There will be a day when God chooses to reveal His Creative ways... perhaps it will include the secret of making something out of nothing and I can't wait!

Closing

The greatest commandment that God gives us is to Love the Lord our God with all our heart, all our souls, all of our mind, and all of our strength! Life's distractions can quickly capture our hearts and find their way into the very core of our being. My prayer is that *Before the Beginning* will nurture a desire to really get to know our Maker and Designer. He is there for you and me . . . personally! God quietly waits for us to choose to be His friend. He desires that His Spirit will flow through us so we can be used in ways we can *never* imagine on our own.

The Renewing of Hearts and Minds

> *"And the peace of God, which transcends all understanding, will guard your hearts and your minds in Christ Jesus."*
>
> Philippians 4:7

As I end this book I want to spend a few minutes on loving God with all of our hearts and all of our minds. Of great importance to the Lord is for us to guard our hearts. Your heart condition is valued by God, and it should be valued by you!

> *"I heard a voice which said, 'There is one, even Christ Jesus, that can speak to thy condition', and when I heard it, my heart did leap for joy."*
>
> George Fox, Founder of Friends Church

Through His Word, God is renewing my mind daily. I am asking God to teach me, to show me more of what I need to know. I can only share with you what I know at this time and place on life's journey.

God isn't finished with me yet and my journey has not ended. He will continue to mold me into His image and transform my mind and my heart into the likeness of Christ. This book is a snapshot, if you will, of how I see my Maker and eternity at this point in my life. My hope is that it will help you along in yours.

> *"In my travels I have found that those who keep Heaven in view remain serene and cheerful in the darkest day. If the glories of Heaven were more real to us, if we lived less for material things and more for things eternal and spiritual, we would be less easily disturbed by this present life."*
>
> Billy Graham, American Clergyman

In chapter 12, I talked about my son's need for going deeper with his faith and His need to see God work in his daily life. Well, that is being addressed ever so conveniently, or should I say, miraculously, as I finish this book. My son is seeing us pray and seeing the Lord provide while we do this work. The Lord is redeeming us by answering prayer, guarding my son's heart, my daughter's heart, my wife's heart and in God's own special ways, He has personally ministered to my heart.

Now that I am in the process of answering His call, I can proclaim in confidence that God is a redeemer! And even as I write this closing, it's not an easy time, but my family is still seeing miracle after miracle. We are now looking forward to seeing God work by getting this book into the "right" hands.

If you can recognize, acknowledge and *accept* God's desire to love each one of us through His Son Jesus Christ *and His* creation, you will see God's epic passion begin on a personal level.

When you see God's creative touch on your life, your loved ones' lives, and all that He put into place to perfectly care for us all, your perspective changes. With a new perspective, we can more easily accept the ups and downs of life. It reminds us that God is in control and He will walk beside us through the barren places as well as the fertile ones.

> When you see God's creative touch on your life, your loved ones' lives, and all that He put into place to perfectly care for us all, your perspective changes.

We can accept other's differences because God made them, loves them and forgives them. Knowing we are not yet perfected in Christ, we can accept others just as God accepts us the way we are.

In 1965, Paul Harvey, a conservative American radio broadcaster, gave an address called, "If I Were the Devil." Mr. Harvey stated that the devil is actively seizing our hearts, our very being. Search in YouTube: "If I Were the Devil", and see if you have fallen into the "do as you please" attitude toward life. You will see that most of Mr. Harvey's contrast-creating words still ring true today. Our nation has been drifting away from the things of God for a long time. However, many are turning to God and seeing our Creator care for their hearts as well.

> *"When you know what God has done for you, the power and the tyranny of sin is gone and the radiant, unspeakable emancipation of the indwelling Christ has come."*
>
> Oswald Chambers

If we can consider what it took in creativity, sensitivity, forethought and transcendent genius in design and planning to create the world around us, we will find hope and a more *eternal perspective*. Understanding the love God demonstrated to us in His provisional design of creation and His personal interest in our hearts can give us encouragement and comfort. We CAN experience awe and wonder on this side of Heaven, reflecting on everything as it was intended to be, *before* the beginning.

Addendum

In its current state, the peer review process can be the birthplace of obstacles that keep us from experiencing God's passion. Each evolutionary or scientific claim goes through a peer review process. Many times it's successful, but other times it's not. If a scientific claim had to go through a US court of law, it would likely fail. According to the American Physiological Society, there are many weaknesses in the system that make it possible to generate misleading information in the area of professional peer reviews:

- No formal training is required for the reviewer. The reviewer learns on the job.
- Unfounded bias in peer and job pressures. If you give a bad review, it is likely that the reviewer won't be asked again. This can lead to loss of objectivity.
- Reviewing is like any other skill; you get better with practice.
- It's hard to find experts on certain subjects. Therefore it is easy to make mistakes.
- School textbook writers and researchers pick up these articles and publishers print them.
- Intentionally delayed "stonewalled" reviews exist because the reviewer is trying to get published first on the subject.
- Scientific merit is compromised when political interest, business or potential sales outcomes of books and magazines influence the reviewer's opinion.
- Uncovering corruption/scientific misconduct is difficult at best.
- There are biases toward certain authors.
- Trust has been damaged and weakened in the scientific community's review process.[1]

If the *scientific community* sees that its own monitoring process is faulty, it stands to reason that some of the outcomes and scientific claims of such cannot (and should not) be fully trusted in academia or industry. At the very least, added scrutiny should be applied.

Just as there should be more careful analysis of information that we take in, we need to simultaneously scrutinize information we distribute. The quality of validation depends on many variables, such as a subject's importance or relevancy, who may be positively or negatively affected by this information, and what would happen to these people if our information is not correct?

Our output needs to carry integrity. If we are presenting information, we are to make every effort to ensure accuracy even if it hurts. We need to slow down and work towards achieving objectivity, in spite of the fact that we are, indeed, human.

Recently I experienced a "divine" meeting while I was dining at a breakfast café. I wasn't seeking this out, but God seemed to have "scheduled" this appointment for me on my calendar. From the booth next to me, I happened to overhear that the gentleman was a teacher and an author. When he got up to leave, I asked him what he taught. He said that he was a professor at one of the local universities and that he taught anatomy and physiology. I told him that I was writing a book and one of the chapters that I was working on happened to be about peer-reviewed information. He knew exactly where I was going and jumped in with his recent textbook error catching experiences. He went on to say that the textbooks recently acquired are filled with errors! It was very frustrating to him to be teaching one thing while the textbook is stating something else. This frustration is undoubtedly facing many educators at all levels of education throughout our nation!

How do I know it was a God appointed meeting? The above paragraph was needed to help make a point. The last paragraph I wrote prior to meeting him was the one above it! I have to mention that many of the other parts of this book happened outside of my own research efforts; all of these elements seemed to be a gentle nudging from the Holy Spirit for me to continue my work in this area.

> Colleges would do well to offer a degree or certification in administration of the peer review process.

It is possible that correcting evolutionary theories may change in our school textbooks as evolutionary proponents literally die off. Our society should make these changes in a more concise and ordered manner than this. Colleges would do well to offer a degree or certification in administration of the peer review process. The peer review process should also reflect what our courts use to verify information. Perhaps a few rules from the Federal Rules of Evidence could be included. Here are four that would be helpful:

Rule # 614	... All parties are entitled to cross-examine witnesses thus called.
Rule # 702	... May testify thereto in the form of an opinion or otherwise, if (1) the testimony is based upon sufficient facts or data, (2) the testimony is the product of reliable principles and methods, and (3) the witness has applied the principles and methods reliably to the facts of the case.
Rule #703	... Facts or data that are otherwise inadmissible shall not be disclosed to the jury by the proponent of the opinion or inference unless the court determines that their probative value in assisting the jury to evaluate the expert's opinion substantially outweighs their prejudicial effect.
Rule #705	... The expert may in any event be required to disclose the underlying facts or data on cross examination.[2]

The peer review process is under increased pressure as scientific journals burgeon due to added research funding. In many instances, editors are no longer experts and US federal agency has loosened peer review rules.[3] This dynamic will only lead to additional unfounded or dated science lingering in our textbooks.

Another method of filtering our information is to remember that the Bible helps to teach us how to manage and filter our thinking.

Notes

Preface

1. "A select list of Science Academics, Scientists, and Scholars Who are Skeptical of Darwinism," JerryBergmanPhD.com, Accessed October 6, 2012, http://jerrybergmanphd.com/articles/?p=94.
2. "Physicians and Surgeons Who Dissent from Darwinism list, Physicians and Surgeons for Scientific Integrity," Accessed October 6, 2012, http://www.pssiinternational.com.
3. Dr. Heribert Nilson, professor at Lund University, in his book *Synthetische Artbildung* (The Synthetic Formation of Kinds) states, ". . . the theory of evolution is a severe obstacle for biological research. As many examples show, it actually prevents the drawing of logical conclusions from one set of experimental material. Because everything must be bent to fit this speculative theory, an exact biology cannot develop."

I: What Was He Thinking?

1. I Corinthians 2: 8–16, New American Standard (NAS).
2. John 15: 15, (NAS).
3. Dr. Werner Gitt, *In the beginning was information*, [sic], (Master Books, Green Forest, AZ, 2007).
4. Richard A. Swenson M.D., *More Than Meets The Eye*, (Nav Press, Colorado Springs, CO, 2000), 73.
5. I Peter 1: 6, 7, (NAS).
6. Philip Yancey and Dr. Paul Brand, *In the Likeness of God*, (Zondervan, Grand Rapids, MI, 2004), 349.
7. Yancey and Brand, *In the Likeness of God*, 353.
8. Swenson, *More Than Meets The Eye*, 38.
9. Swenson, *More Than Meets The Eye*, 39.
10. Swenson, *More Than Meets The Eye*, 40.
11. The Free Dictionary, "Passion," accessed Oct. 6, 2012, http://www.thefreedictionary.com/passion.
12. Mark Batterson, *Primal*, (Multnomah Books, New York, NY, 2010), 7.

13. Sam A. Smith, "Who Wrote 'the Psalms of David?'", 2012, *The Biblical Reader*, Accessed Oct. 8, 2012, http://www.biblicalreader.com/btr/Who_Wrote_the_Psalms_of_David.htm.
14. Isaiah 42: 21, I Corinthians 1: 21.
15. ABC News, "Terri Irwin's *20/20* interview with Barbara Walters", September 27, 2006.
16. Dr. Jobe Martin, *Incredible Creatures that Defy Evolution I*, DVD.

2: God's Heart In Communication With Man's Heart

1. Hans Walter Wolff, *Anthropology of the Old Testament* (Fortress Press, Philadelphia, PA, 1975), 40.
2. Luke 8: 14, (NAS).

3: Inspirations of God

1. Science, Discovery Channel "The Human Body: Pushing the Limits," Episode aired on October 8, 2012.
2. Dr. Jason Lisle, *Taking back Astronomy, The Heavens Declare Creation*, (Master Books, Green Forest, AZ, 2006), 33, AiG Bookstore: http://www.answersingenesis.org/store/.
3. Encyclopedia Britanica, "Strong Force," a fundamental interaction of nature that acts between subatomic particles of matter. The strong force binds quarks together in clusters to make more-familiar subatomic particles, such as protons and neutrons. It also holds together the atomic nucleus and underlies interactions between all particles containing quarks. Accessed October 8, 2012, http://www.britannica.com/EBchecked/topic/569442/strong-force.
4. Encyclopedia Britannica, "Golden ratio," also known as the golden section, golden mean, or divine proportion, in mathematics, the irrational number $(1 + \sqrt{5})/2$, often denoted by the Greek letters τ or ϕ, and approximately equal to 1.618. Accessed October 8, 2012, http://www.britannica.com/search?query=golden+ratio.
5. Genesis 6: 5–6, Hosea 11: 7–8, Ephesians 4: 30–31, (NAS).
6. Matt 13: 22, 23, (NAS).

4: God's Creation: Get the Message

1. A photographer's term used to describe a quality of light that can only occur at sunrise as well as late in the day around sunset. The sun's angle is such that the rays travel through a higher concentration of atmospheric particles, adding either the soft pastel look or a glowing quality.

2. Swenson, *More Than Meets The Eye*, 37.
3. CBS News, "Deep Sea Volcanic Vent May Offer Discoveries" Accessed October 8, 2012, http://www.cbsnews.com/2100-205_162-6388402.html.
4. Alaska Public Lands Information Centers, "Ice Worms", Accessed October 8, 2012, http://www.alaskacenters.gov/ice-worms.cfm.
5. Lisle, *Taking back Astronomy, The Heavens Declare Creation*, 17.
6. Lisle, *Taking back Astronomy, The Heavens Declare Creation*, 16.
7. Baskin Robbins is a registered trademark of BR IP Holder LLC.
8. Paul Garner, *The New Creationism*, (EP Books, Carlisle, PA, 2009), 134.
9. Dr. Jobe Martin, *Incredible Creatures that Defy Evolution III*, DVD.
10. Gitt, *In the beginning was information*, 18

5: Diversity and Devotion

1. Ecclesiastes 12: 13, (NAS).
2. Psalm 139, (NAS).

6: The Garden Around Us

1. Swenson, *More Than Meets The Eye*, 38.
2. Acts 17; 24–28, Job 12: 10, Hebrews 1: 3, (NAS).
3. Ephesians 1: 4, (NAS).
4. I Timothy 6: 17, (NAS).

7: 1.2 oz.–2,000 lb. Beasts

1. McDonald's is a registered trademark of McDonald's Corporation.
2. TCBY is a registered trademark of Mrs. Fields Famous Brands: Dairy Queen is a registered trademark of AM.D.Q. CORP. Ben and Jerry's is a registered trademark of *Ben & Jerry's* Homemade, Inc.
3. Outback is a registered trademark of Bloomin' Brands, Inc. LoneStar is a registered trademark of Lone Star Texas Grill.
4. Jim Carrey interview, "Inside the Actors Studio," hosted by James Lipton, season 17, episode 2, original airdate: 2011-01-10.
5. Vance Ferrell, *Marvel of God's Creation #1* (Harvestime Books, Altamont, TN, no publishing date offered).
6. CBSNEWS, "Deep Sea Volcanic Vent May Offer Discoveries," April 12, 2010, http://www.cbsnews.com/2100-205_162-6388402.html.
7. Ferrell, *Marvel of God's Creation #6*.
8. Ferrell, *Marvel of God's Creation #10*.

9. Malleefowl Preservation Group, Inc., "The Bird," Accessed September 17, 2012, http://www.malleefowl.com.au/?page_id=29.

8: The Masters and Masterpieces

1. J.K. Rowling stated that one day she would be "found out" in a recorded interview at the Roald *Dahl* museum in Buckinghamshire, United Kingdom. Visited in July, 2005.
2. Michael J. Gelb, *How To Think Like Leonardo da Vinci*, (Dell Publishing Group, New York, NY, 2000), http://www.michaelgelb.com.
3. Michael J. Gelb, *How To Think Like Leonardo da Vinci*, 47.
4. Philippians 3: 13, (NAS).
5. John 15: 5, Galatians 5: 1, (NAS).
6. II Corinthians 5: 17, (NAS).
7. John 8: 32, Psalm 51: 12, (NAS).
8. II Corinthians 3: 18, (NAS).
9. Richard Deem, Evidence For God, "Did Albert Einstein Believe in a Personal God?", Accessed October 10, 2012, http://www.godandscience.org/apologetics/einstein.html.
10. Daniel S. Burt, *The biography book: a reader's guide to nonfiction, fictional, and film biographies of more than 500 of the most fascinating individuals of all time.* (Greenwood Publishing Group, 2001), 315.
11. James R. Graham, The Early Period (1608–1672), 250.
12. Webb, R.K. ed. Knud Haakonssen, "The emergence of Rational Dissent." Enlightenment and Religion: Rational Dissent in eighteenth-century Britain, (Cambridge University Press, Cambridge, England, 1996), 19.
13. Gitt, *In the beginning was information*, [sic], 105.
14. Austin Brown, The Long Foundation, "Long Term Art," *The Lost (and Found?) Battle of Anghiari*, states, ". . . a story that could likely be behind the hiding of the Fresco. The painting was experimental as DaVinci was working on a new paint pigment." Published May 16, 2012, http://blog.longnow.org/02012/05/16/the-lost-and-found-battle-of-anghiari/.
15. Yancey and Brand, *In the Likeness of God*, 533.
16. Brown, The Long Foundation, "Long Term Art", *The Lost (and Found?) Battle of Anghiari*, Notation found in paragraph 9.

9: Unveiling Creation Design

1. National Geographic, "Brain Games", Accessed October 10, 2012, http://channel.nationalgeographic.com/channel/brain-games/videos/the-switcheroo/.
2. Deuteronomy 11: 19, Ephesians 6: 4, Deuteronomy 4: 9.

10: Obstacles That Keep Us From God's Passion

1. Jeffrey Tomkins, Ph.D., Institute for Creation Research, "Gene Control Regions Are Protected—Negating Evolution", Accessed October 10, 2012, www.icr.org/article/6886/.
2. YouTube, "Antony Flew's conversion to theism," Accessed September 20, 2012, http://www.youtube.com/watch?v=X1e4FUhfHiU&feature=related.
3. Garner, *The New Creationism*, 13.
4. Chuck Missler, The Myths of Science, *Challenging the Myths of Astronomy in The Electric Universe*, Accessed October 10, 2012, http://www.youtube.com/watch?v=CvDqrSTCcmA, 26 minutes in on the video.
5. Chuck Missler, The Myths of Science, *Challenging the Myths of Astronomy in The Electric Universe*, Accessed October 10, 2012, http://www.youtube.com/watch?v=CvDqrSTCcmA.
6. Tomkins, "Gene Control Regions Are Protected—Negating Evolution."
7. Mike Riddle, *The Riddle of Origins Series*, DVD series, "Answers In Genesis," Petersburg, KY 2005.
8. *Forbes*, "Climategate 2.0: New E-Mails Rock The Global Warming Debate," Accessed October 10, 2012, http://www.forbes.com/sites/jamestaylor/2011/11/23/climategate-2-0-new-e-mails-rock-the-global-warming-debate/.
9. Swenson, *More Than Meets The Eye*, 162.
10. Ken Ham and Britt Beemer with Todd Hillard, *Already Gone*, (Master Books, Green Forest, AZ, 2009), 81.
11. Dr. Stuart Firestein, *Ignorance: How it Drives Science*, (Oxford University Press, New York, NY, 2012), no page offered.
12. Rodney Stark and Byron Johnson, *Religion and the Bad News Bearers*, Wall Street Journal article, 8/26/2011, Accessed October 10, 2012. http://online.wsj.com/article/SB10001424053111903480904576510692691734916.html.
13. Ham, Beemer and Hillard, *Already Gone*, 73.
14. Pete Briscoe with Todd Hillard, The Surge, "Six Graphs that will Change Your View of the World . . . and Life.", Accessed October 10, 2012, http://getinthesurge.com/?page_id=143.

15. Hebrews 5: 12–14, (NAS).
16. "Physicians and Surgeons Who Dissent from Darwinism list," Physicians and Surgeons for Scientific Integrity, Accessed October 6, 2012, http://www.pssiinternational.com.
17. Peter Moore, "Inferential Focus Briefing," September 30, 1997.
18. YouTube, "Statistics," Accessed September 19, 2012, http://www.youtube.com/t/press_statistics.

11: Transformers

1. Mathew 5: 45, (NAS).
2. I Corinthians 15: 51, II Timothy 3: 16, 17, Deuteronomy 31: 6, (NAS).
3. Garner, *The New Creationism*, 54.
4. "... We are of good courage, I say, and prefer rather to be absent from the body and to be at home with the Lord." (II Cor. 5: 8), "Not everyone who says to Me, 'Lord, Lord,' will enter the kingdom of heaven, but he who does the will of My Father who is in heaven will enter.", (Matt. 7: 21), [sic].
5. James 1: 5, (NAS).
6. YouTube, "Lee Strobel Testimony," Accessed September 20, 2012, http://www.youtube.com/watch?v=2AT_bMuFBfs.
7. YouTube, "Josh McDowell Testimony Part 1," Accessed September 20, 2012, http://www.youtube.com/watch?v=d5O5nD0pyPc.
8. YouTube, "Antony Flew's conversion to theism," http://www.youtube.com/watch?v=X1e4FUhfHiU&feature=related.
9. Hebrews 3: 9, (NAS).
10. YouTube, "Ted.com talk—Conception to birth visualized—Alexander Tsiaras," Accessed September 22, 2012, http://www.ted.com/talks/alexander_tsiaras_conception_to_birth_visualized.html.

12: Hope in the Passion Creator

1. John 15: 5, (NAS).
2. I Corinthians 2: 9, (NAS).
3. James Strong, *Strong's Exhaustive Concordance of the Bible*, (Dugan Publishers, Inc., Gordonville, TN).

Addendum

1. J Educ Eval Health Prof. 2008; 5: 5. Published online 2008 December 22.
   A Personal View, *ADV PHYSIOL EDUC 27:47–52, 2003* © 2003 American Physiological Society, Published online 26 July 2011 | Nature doi:10.1038/news.2011.441.
2. Federal Rules of Evidence, December 1, 2009.
3. Jeff Tollefson, "US federal agency loosens peer-review rules," Nature, July 26, 2011, doi:10.1038/news.2011.441.

Author Biography

Jim Kraft grew up in Willoughby, Ohio—a suburb of Cleveland. Always a true designer (small d) at heart, he often spent time at his father's business designing and creating "prototype" projects. Later, Jim earned his diploma from the Cleveland-based Cooper School of Art & Design and graduated with honors.

After freelancing for a few art studios and ad agencies, he was hired by a Cleveland ad agency as a designer and art director. For most of his career, Jim has been an independent design consultant creating on and off-line communications for major corporations like GE Lighting, Petro Canada, American Greetings, BF Goodrich, Allstate Insurance, Carlton Cards, Honeywell, Holiday Inn, and Dutch Boy. He has also done work for regional organizations like Parkside Church, Friends Church-Willoughby Hills, EZPOLE Flagpoles and Maple Valley Sugarbush & Farm. He has performed marketing communication services for over 70 church, Para-church and secular organizations.

Jim's mid-life crisis didn't include an attention-getting hot car but instead, a small herd of lowly cows (Black Angus to be specific). Jim says that the ride is "a little bumpy, but you get just as many looks!"

Jim has always felt that God has given him the gift of helping people to "see the light" and (in his spare time) represents a natural light manufacturer. It's not quite what he had in mind for earning a living, but God has used it to help him get by while he wrote *Before the Beginning*.

Jim and his wife Tammy reside in Waite Hill, Ohio and have two children. Jim has taught marriage enrichment and biblical creation courses and has a true passion for church growth. He enjoys serving on the outreach committee in his congregation and horsing around at his family farm.

Endorsements

> **Bob Devine**
> MBI Nature Corner Radio Broadcaster
>
> Faithfully hosted radio program for WCRF Cleveland, Ohio, for over 30 years. Produced 490 episodes of interviews about various wonders of science. Affectionately known as "Uncle Bob."

"The highest known mountain in the world is 29,035 foot Mount Everest located on the borders of Nepal and Tibet. It's the Behemoth of the Himalayas. How old might it be? Moses, that great Jewish Old Testament General writes in the 90th Psalm, verse 2, 'LORD, BEFORE the mountains were born or you brought forth the earth and the world, from everlasting to everlasting, you are God!' Think about that! Before there was anything in the skies above or oceans beneath, The Eternal God was already there. What was our Universe like *Before The Beginning*? Does that sound like a contradiction? Jim Kraft, author of this book, *Before The Beginning* asks, 'Was God just bursting forth with excitement at the thought of sharing His Divine understanding and Creation to come with His future image-bearers? When love is demonstrated there needs to be an object of our love, or in this case, God's love.' John 3:16 opens with, 'For God so loved the world. . . .' 'This means you and me! God is sharing what He possesses, Eternal Life, and all that accompanies this privilege.' In *Before The Beginning*, Jim Kraft cites both an abundant amount of Scripture and Scientific examples of both truth and science all around us, showing us that wherever there is a *design*, there demands a *Designer*. What is His Name? Colossians I: 13–17 says the Designer is The Son of God, The Lord Jesus Christ. I highly recommend Jim's book, *Before The Beginning* for exciting, truthful reading."

> **Gerald Hillier, M.S Physics**
> Senior Engineer
> Siemens Energy Inc.
>
> - 15 years engineering experience in aerospace and aviation with an emphasis on development of processes and materials benefiting the aerospace, military and energy production markets.
> - Specialty: Applied Optics, Ceramics and Characterization of Materials using laser spectroscopy.

"This recent work by Jim Kraft treats the Creation of God and the relationship of God to humanity via its perception of creation, as a potential issue in a way most cannot verbalize. Jim gives a logical approach to understanding a complex issue by first acknowledging reverence to the creative genius of God with many untypical, concrete examples. Second, potential motives of creation, (relative and confined to human understanding), are discussed as a means to discover God's love for humanity with the clear hope of facilitating deeper relationship with the Ultimate Designer, God of the Universe.

When applicable, examples from modern biology, astronomy, and physics are used as supportive evidence of merging creative genius and motive with God's desire for relationship; all of this is in combination with scripture relevance. However, this is not a science text, and the author does not make recurring attempts or claims to convince the reader with empirical evidence or existing counter-arguments. Rather, the reader is urged to view the creation of the universe with open eyes and open heart in an attempt to draw near to its ultimate Designer. The examples need not be put on trial as they are presented to be divinely self-evident. The tone here is not of conflict with science, but of reconciliation. For those simply looking for a re-statement or re-building of their faith, this book provides many examples, relevant, personal ideas, and relevant scriptures."

> **Ron DiCianni**
> Author and Painter
>
> Recognized as one of the nations top illustrators. In 1989, he produced the painting that started a revolution . . . *Spiritual Warfare*. No stranger to the book industry, Ron has collaborated on over 50 book projects, is a six-time winner of the Gold Medallion Award for Excellence in Christian Literature and Retailers Choice Book of The Year winner with Randy Alcorn. These projects have generated multiple millions of copies in print and have been translated into languages all over the globe. Some of his work can be viewed at www.TapestryProductions.com.

"A story I heard illustrates the battle written about in this book very well. A group of scientists challenged God to a contest. They felt they could create a human being just as well as He could. So the contest began. God did what He does and created a perfect human being. Embarking on their attempt, they started with the same mud God did, whereupon God interrupted them and said, *"Go get your own mud . . ."* That's the story in a nutshell, isn't it? God either is or He isn't. And if He is, our attempts to erase Him, preclude Him or override Him will never work. Someday it will become much clearer than it is right now, but by then it will be too late to recant. This book may help you come to the truth of God before that day, for the truth, dear reader, is that it doesn't matter what you believe. It's only matters what is true."

> **Kay Kyllonen, Pharm.D., FPPAG,**
> Neonatal ICU Clinical Pharmacy Specialist
>
> Clinical Pharmacist in Pediatrics, mother of two teenagers.

"I am enjoying this book. I have been reading it with my Bible. I love the integration of Scripture, science (which in its truest form never contradicts the word of God) and concepts of design you use to enliven the reader's perception of God. I will be recommending it to others as an engaging look at the study of God's creation. Thanks for giving me the privilege to review your (and His) handiwork."

The heart of it.

Man can create great works of art that capture man's attention and cause wonder and appreciation. We call these men and women "Masters," and their creations, masterpieces. However, these masterpieces do not compare with God's masterpieces. He *is* the Master Designer. One of God's masterpieces is within each of us. It is our hearts. He authored, designed and engineered our hearts with two distinct purposes. One purpose is to power our bodies and the other is to possess a heart capable of emotion, of passion, sadness, joy, anger and, above all else, love.
Unfortunately, tumultuous times can skew our perspective and affect the condition of our hearts. Whether young or old, we can begin to *devalue* the miracle of who we were meant to be. *Before the Beginning* addresses the emotional, physical and spiritual fallout of a lost perspective and offers the healing and enduring hope of *God's perspective* through His *word* and the *creation* that He put before us.
Our hearts are God's masterpieces. They are so important to Him that He tells us that "above all else, guard your hearts." When we guard our hearts we can more easily grasp our Creator's heart. Our heart connects us to God. If that connection changes us, we become more like Christ. We slowly begin to understand the incredible love and passion that God has for us.
Who else would be better to address *heart issues* other than God, the Artist, Author, and Designer of our hearts?